Computer Communications and Networks

The **Computer Communications and Networks** series is a range of textbooks, monographs and handbooks. It sets out to provide students, researchers, and non-specialists alike with a sure grounding in current knowledge, together with comprehensible access to the latest developments in computer communications and networking.

Emphasis is placed on clear and explanatory styles that support a tutorial approach, so that even the most complex of topics is presented in a lucid and intelligible manner.

More information about this series at http://www.springer.com/series/4198

K. Erciyes

Distributed Real-Time Systems

Theory and Practice

 Springer

K. Erciyes
Department of Computer Engineering
Üsküdar University
Üsküdar, Istanbul, Turkey

ISSN 1617-7975   ISSN 2197-8433   (electronic)
Computer Communications and Networks
ISBN 978-3-030-22572-8   ISBN 978-3-030-22570-4   (eBook)
https://doi.org/10.1007/978-3-030-22570-4

This Springer imprint is published by the registered company Springer Nature Switzerland AG
The registered company address is: Gewerbestrasse 11, 6330 Cham, Switzerland

To the memory of Professor Sinan Yılmaz who was a keen educator and a researcher on real-time systems, and to Professor Nevzat Tarhan for his strong belief and commitment in scientific research and the good things that naturally follow

Preface

Distributed and embedded real-time systems are everywhere from factory floor to modern cars to avionics. A distributed real-time system is characterized by a number of computational nodes that are connected by a real-time network and cooperate to finish real-time tasks. A real-time task has a deadline and finishing of the task before its deadline is imperative in many applications. Recent technological advancements have provided magnitudes of increase in the number of nodes of such distributed real-time systems which has resulted in a tremendous need for the software design of such systems. A distributed real-time system node has some computational capacity and is commonly interfaced to external world using sensors and actuators. Not all embedded systems are real time and we will use the term distributed real-time system to cover the distributed embedded systems that have real-time characteristics.

This book is about the design and implementation of software for distributed real-time systems using a bottom-up approach. Having taught related courses at undergraduate and graduate levels for few decades, and being involved in large software projects associated with real-time systems, I have been able to observe the main bottlenecks for the designer and the implementer. First of all, the designer/implementer is often confronted with the challenge of interfacing the application with some commercial real-time operating system/middleware and sometimes writing patches to it. This requires an in-depth understanding of the hardware and operating system concepts related to real-time processing and we have a part related to system software for this reason. This part has three chapters in the first chapter of which we review basic real-time operating system concepts. We then construct an experimental distributed real-time operating system kernel (DRTK) from scratch in the second chapter by showing all necessary details. Consequently, we review distributed real-time operating system and middleware concepts and describe how to design network communications so that the real-time kernels can cooperate to have a distributed real-time system software frame in the last chapter. The experimental kernel is gradually transformed into a kernel of a distributed real-time operating system with associated middleware, showing all details of implementation in the proceeding parts of the book.

The second challenge is of course scheduling of tasks so that all meet their deadlines. Tasks in a real-time system can be broadly classified as hard, soft, and firm real-time tasks which can be periodic and aperiodic requiring different scheduling strategies. Moreover, they can be independent or dependent on each other necessitating means for inter-task synchronization in the latter. Yet, we need to provide end-to-end scheduling of tasks so that deadlines are met and load is distributed evenly to the nodes of the distributed real-time system. Another related problem is the management of resources over the network and all of these problems are investigated in Part II, always with implementation in mind.

Finally, the designer is faced with the task of performing all of the steps of software engineering starting with specifying the requirements, then the high-level design, detailed design and coding. This process turns out to be where most difficulties are encountered. We provide a novel method to perform all of these steps simply and effectively. We have a chapter on high-level design and detailed design using finite-state machines which can be implemented using operating system threads. Fault tolerance is imperative in a real-time system to prevent catastrophic events and we have a chapter on this topic. We also review various real-time programming languages including C/POSIX, Ada, and Java. Finally, a real-time case study is presented in the final chapter by incorporating all of the methods we have developed and reviewed starting with high-level design followed by detailed design and then coding.

In each chapter, we first review the concepts and then describe methods to design and implement the needed software by giving a brief survey of commercially available software. In the second part of a chapter where applicable, we show how to implement the described concepts in the experimental sample kernel by displaying the executable code. These parts are commonly called "DRTK Implementation" in each chapter and can be skipped along with Chap. 5 which describes DRTK, when the book is used in a course with limited coverage of real-time systems. Finally, we provide review questions and then a summary of the chapter by emphasizing main points, give a contemporary review of the related literature, and show possible open research areas in the Chapter Notes section.

About DRTK

The main modules of DRTK described in detail in Chap. 5 are tested during teaching of various courses related to real-time processing. However, the parts related to distributed processing in the chapters that follow this chapter are not fully tested which means they may, or more likely, they would have some syntax or other implementation errors. I will keep a homepage of the book at http://akademik.ube.ege.edu.tr/~erciyes/DRTS for the code of DRTK, teaching slides and errata. Any modifications to the code of DRTK are welcome with the hope of making it a practical, experimental distributed real-time kernel that can be studied in related courses.

The intended audience for this book are the senior undergraduate and graduate students and researchers of electrical and computer engineering, computer science, engineering in general, or any person with basic background in computer architecture and operating systems. The text contains considerable size of C code for the sample DRTK implementations and various examples. I would like to thank undergraduate and graduate students at various universities I have taught Real-time Systems, Embedded Systems, Advanced Operating Systems or similar courses including Ege University, Oregon State University, University of California Davis, California State University San Marcos, Izmir Institute of Technology, Izmir University, and Üsküdar University in chronological order, for their valuable feedback when various parts of the material covered in the book were presented during lectures and the sample kernel was experimented in laboratory work. I would also like to thank Springer Senior Editor Wayne Wheeler and Associate Editor Simon Rees for their continuous support during the course of writing the book.

Istanbul, Turkey   K. Erciyes

Contents

Part I
Introduction

Introduction to Real-Time Systems

1

1.1 Introduction

Real-time systems are everywhere, from cars and cell phones to avionics to nuclear plant control. A real-time system is characterized by its timeliness in response to some input. Failure to respond in a specified time may result in catastrophic failures in these systems. The correctness of a real-time system depends both on the correctness of the results and the time these are produced which are termed *logical correctness* and *temporal correctness* consecutively. In an *embedded real-time system*, the real-time computing element is embedded in the system to be controlled and most of the real-time systems fall into this category.

There are many types of real-time systems; a real-time process control system receives input data from sensors, performs some operation on this data, and produces output to control various system functions such as turning switches on and off, activating alarms if necessary, and display system data. An aircraft is a real-time system with strict time constraints to be met whereas failing to meet the time constraints of a multimedia system will generally result in poor performance only. Other examples of real-time systems include robotic systems, nuclear power plants, and mobile phones.

We review basic concepts related to real-time systems in this chapter starting with definitions, types of real-time systems, and properties of these systems. We then describe the operation of exemplary real-time systems and conclude by outlining the contents of the book.

© Springer Nature Switzerland AG 2019   3
K. Erciyes, *Distributed Real-Time Systems*, Computer Communications
and Networks, https://doi.org/10.1007/978-3-030-22570-4_1

1.2   What Is a Real-Time System

A real-time system operates with a time constraint, where the time that the output produced is significant. In other words, producing the output at some other time than the required time may be meaningless.

Alternatively, a real-time system is a data processing system which responds to external inputs within a finite specified time called *deadline*. Failure to respond to the input within a deadline may cause harm to lives and property. A real-time system is designed according to the dynamics of the physical processes it has to control. A real-time computing system consists of hardware and software working with time constraints. The real-time software embodies the following components in general:

- *A Real-time Operating System*: An operating system has two main functions: providing convenient access to the hardware and efficient management of resources. For this purpose, it provides process (task) management, memory management, and input/output management in general. A real-time operating system still has these tasks to perform but with strict time constraints. A real-time operating system is typically small in size to be easily accommodated in embedded systems and it has to be fast with minimum operating system overheads.
- *A Real-time Programming Language*: A real-time programming language provides basic schemes such as inter-task communication and synchronization, error handling, and scheduling for real-time tasks. Note that these functions are also provided by the real-time operating system. Hence, a real-time programming language can be used alternatively to a real-time operating system to achieve real-time processing. At the lowest level, an assembler can be used to access registers and other hardware of the processor. However, an assembler language is error prone and is not portable to another processor. The C language is widely used in real-time systems as it provides access to hardware and has very simple input/output interface. However, various other languages such as Ada and real-time Java are developed for real-time systems to ease their programming.
- *A Real-time Network*: A computer network provides transfer of data between various computing devices. A real-time network should provide a reliable delivery of messages within specified time limits. Key to the operation of a real-time network is the reliable and timely transfer of messages. A real-time communication protocol provides timely and guaranteed delivery of messages over the network.

At this point, it may be right to distinguish a real-time system from a system that looks real time but in fact does not conform to what is generally accepted as real time. For example, an *online system* interfaced to a user is not considered real time since tasks have no defined deadlines. Also, a system that has to react *fast* does not necessarily mean it is real time.

An *embedded system* is characterized by the computing element being an element contained in the system. Embedded systems are not general computing systems where the computer may be programmed dynamically to achieve different tasks. On the contrary, they are built for a special purpose such as a microwave oven. An embedded system is tightly coupled to the external physical world. Home and personal appliances that are used to assist living are examples of embedded systems. In fact, embedded systems constitute majority of computer-based systems. An embedded system may be real time or not depending on the application. A non-real-time embedded system has no time bounds, for example, an MP3 player is a non-real-time embedded system. However, we find significant percentage of real-time applications is embedded. An embedded real-time system is designed for a specific function and works with real-time constraints. Our terminology in this book refers to real-time systems in general, including real-time embedded systems.

Real-time systems can be classified using various approaches; one main arrangement is based on the synchronization and interaction among various components of the system. Consequently, a real-time system may be time driven in which each activity has a specific time assigned to it; event driven where external inputs determine the behavior of the system or interactive where the user may change the mode of the operation. In many cases, a real-time system works as a combination of all of these modes.

1.3   Basic Architecture

A real-time system consists of the following components in hardware as shown in Fig. 1.1, which is detailed as follows:

- *Sensors*: A sensor converts a physical parameter into an electrical signal to be processed. For example, a thermocouple is a heat sensor that converts heat to an electrical signal.
- *Processing Element* (PE): The real-time processing element or the real-time computer receives digital data as provided by the input interface unit, processes data, and decides on some output function to perform.
- *Actuators*: An actuator inputs the electrical signal from the output interface of the processing element and actuates a physical action. A relay is an example of an actuator that opens/closes a switch when activated by a signal.
- *Input Interface*: The input interface to the real-time computer processes and converts the electrical signal to digital form to be processed. Sensors may have the needed converters embedded in them.
- *Output Interface*: The binary output from the real-time computer is processed and converted to the form of signal needed by the actuator or other output devices in the output interface.

Fig. 1.1 Architecture of a typical real-time system

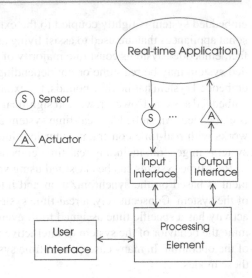

1.4 Characteristics of Real-Time Systems

Common characteristics of real-time systems are as follows:

- *Meeting Deadlines*: Deadlines of all hard real-time tasks must be met to prevent loss of lives and property.
- *Large Size*: Real-time systems are commonly large and complex in terms of hardware occasionally but mostly in the size of the software used. Even a small real-time system may have hundreds of thousands of lines of code to run.
- *Predictability*: We need to be able to predict the worst-case response times and whether the deadlines of all tasks will be met prior to the execution of tasks in a real-time system. Predictability commonly involves proving theoretically that all deadlines will be met in a hard real-time system.
- *Safety and Reliability*: Real-time systems operate and control environments such as nuclear plants and process control, where safety and reliability are of utmost concern. Unsafe and unreliable systems are prone to errors which may result in loss and damage to lives and property.
- *Fault Tolerance*: A *fault* is said to occur when a component of hardware or software of a computer system fails and the system does not behave according to its specification anymore. A *failure* is the result of a fault, which may cause loss of lives and property in a real-time system. A fault may be permanent causing the system to fail permanently or transient, which may disappear after some time. Fault tolerance is the ability of a system to resist faults and continue functioning correctly in the presence of faults.

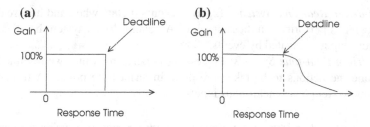

Fig. 1.2 Utilization values of **a** hard and firm real-time systems, **b** soft real-time system

- *Concurrency*: The physical environment to be controlled by the real-time computer commonly exhibits parallel execution of events. The real-time computer must be able to cope with this parallel operation using concurrent system software capabilities or distributed hardware.

1.5   Classification of Real-Time Systems

We can specify the types of real-time systems in terms of the input received, on timing constraints, and the type of processes in the system. We can have the following classification when deadlines are considered:

- *Hard Real time*: Missing a *hard deadline* may cause catastrophic results, which may result in the loss of property or lives. Aircraft maneuvering systems, chemical plants, and nuclear power stations are examples of such systems. All deadlines of tasks are hard in a *hard real-time system*. The gain or the utilization of the system when a deadline is missed is zero in such a system as depicted in Fig. 1.2a.
- *Soft Real time*: Missing a *soft deadline* is not critical, however, system quality of service degrades when this happens. A *soft real-time system* has tasks with soft deadlines. Multimedia systems and airline reservation systems are examples of soft real-time systems. The gain or the utilization of the system when a deadline is missed degrades but does not go to zero immediately as shown in Fig. 1.2b.
- *Firm Real time*: These systems are basically soft real-time systems in which there is no gain when a *firm deadline* is missed. The utility of the system goes down to zero immediately in these systems as shown in Fig. 1.2a but the system tolerates these deadline misses.

A real-time system may have a mixture of tasks that are hard, soft, and firm real time. The processes in the system may be *static* with deterministic characteristics known beforehand whereas characteristics in a dynamic real-time system vary with time. Another distinction is about how processing is performed as below.

- *Event-driven Real-time Systems*: External events dictate when and how real-time processing is performed in these systems. An event-driven system handles asynchronous inputs activated by events.
- *Time-driven Real-time Systems*: Processing is performed in a well-defined time base and the actions to be taken are done in strict time points. A time-driven real-time system has synchronous inputs.

For example, a humidity monitoring of air with a sensor measuring the humidity every t time units is a time-driven system, whereas a system that is activated by an asynchronous external event is an event-driven system. Commonly, we have a mixture of these modes in a typical real-time system. For example, temperature control of an oven can be performed by measuring the temperature at regular intervals (time driven) and turning the heater off when door of the oven is opened (event driven).

A distributed system consists of autonomous computers each of which has the capability to function alone. These computers are connected by a network and cooperate to perform global tasks and share resources. A distributed real-time system is a distributed computing system that works with time constraints.

Distributed systems provide efficient resource sharing and redundancy. Also, distribution of computation across a number of nodes is convenient since fast local processing is possible. Yet another reason to employ a distributed real-time system is that various subsystems by different producers can be interfaced using standard communication protocols. A car is a distributed real-time system since data read from many sensors is transferred over a real-time network to various nodes to be processed. Another advantage gained in using a distributed real-time systems is increased reliability since the failure of a node may not affect the operation of the whole system significantly when precautions are taken in advance.

1.6   An Example System: Bottling Plant

Let us look at a very simple real-time process control system where milk is bottled. The milk bottles move over a conveyor belt and there are three machines; a washer to wash the bottles, a filler that fills milk, and a capper to put the cap on the filled bottle as shown in Fig. 1.3.

Each machine is interfaced to the real-time computer system with sensors and actuators. The computer system has to perform three tasks in sequence as follows:

1. *Washing*: The arrival of a dirty bottle over the conveyor belt is signaled by a sensor. The belt is stopped and the washer is activated for t_w seconds and then the belt is activated. We see here an event-based (sensing the coming of bottle) operation.
2. *Filling*: The coming of bottle is sensed and the filling begins by activating the filler. The level of milk is monitored by a sensor and it stops when the desired level is reached.

Fig. 1.3 Architecture of a
typical real-time system

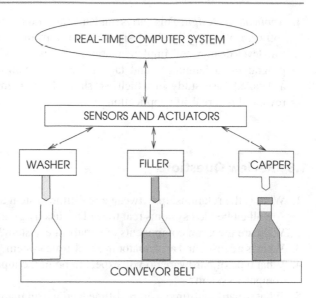

3. *Capping*: When bottle arrives under this machine, a cap is placed on top of it
   and the bottle is ready to be packed.

We see dominantly event-driven processing here. However, with careful timing,
the whole process can be performed time driven by periodically performing the
washing, filling, and capping processes assuming there are always empty bottles to
process. Note that the whole process is slow leaving considerable idle time for the
computer to do other processing.

1.7   Outline of the Book

We have four sections in the book as follows:

1. *Introduction*: This part serves as an introduction to real-time systems with chapters on real-time architecture and distributed real-time systems.
2. *System Software*: The real-time operating system is the focus of this part where
   we describe the basic concepts such as task, memory, and input/output management. We also provide a detailed step-by-step construction of a real-time
   operating system kernel which is used to test various higher level implementations in the proceeding chapters.
3. *Scheduling and Resource Management*: Scheduling of tasks so that deadlines
   are met is a fundamental function of any real-time system. We first describe
   independent periodic and aperiodic scheduling and then dependent scheduling
   with resource management and distributed scheduling in this part.

4. *Application Design*: This part is fundamental to aid designers of real-time system software. We describe the design procedure from high-level design methods to low-level design and implementation. We also provide a survey of real-time programming languages and fault tolerance techniques. This part ends with a detailed case study in which we show the implementation of the methods reviewed in a real-life application.

1.8 Review Questions

1. What is the relationship between a real-time system and an embedded system? Are all embedded systems real time? Discuss by giving examples.
2. What are the main components of a real-time system?
3. What is a sensor and an actuator in a real-time system? Give an example of each.
4. What type of hardware do you expect to be in the input interface of a real-time computer system?
5. What is hard real-time, soft real-time, and a firm real-time system?
6. Give an example each of event-driven and time-driven real-time system.
7. What are the main characteristics of a distributed real-time system?

1.9 Chapter Notes

We have looked at the main components of a real-time computer system, its basic architecture, and different classifications of these systems. Real-time systems can have hard, soft, and firm deadline tasks when tasks are considered. On the other hand, the processing can be handled as event driven and time driven. A real-time computer can be embedded in the environment it has to control in an embedded system and it can be distributed over a real-time network. Generally, a real-time system has a combination of these properties. General concepts about real-time systems are described in [2] and real-time programming languages are contained in [1].

References

1. Burns A, Wellings A (2001) Real time systems and programming languages: Ada 95, Real-time Java and real-time C/POSIX, 3rd edn. Addison-Wesley
2. Laplante PA, Ovaska SJ (2011) Real-time systems design and analysis: tools for the practitioner, 4th edn. Wiley

The Hardware

2.1 Introduction

Most of the real-time computing systems employ general-purpose hardware components such as the processor, memory, and input/output units. An embedded real time system may have special input/output interface units dedicated for the application. The main difference between a non-real-time system and a real-time system at hardware-specific software level is the structure of the low-level software such as handling of the interrupts and time management. In order to fully understand the operation of this software and operating system interface to it, we need to look at the hardware characteristics.

In this chapter, we review the hardware of a real-time system node at component level by deferring the review of the architecture and basic software structure of a distributed real-time system to the next chapter. We start with the basic interface units: sensors and actuators and review characteristics of the common types of these components. The processor is where the actual processing is performed and single cycle, multi-cycle, and pipelined datapaths are discussed next. Memory and input/output units are other main parts of a real-time system as we will see. We will use the MIPS processor [1] to describe the hardware-related concepts as this processor is widely used in embedded systems and yet simple enough to display operation.

2.2 Processor Architecture

The *Von Neumann model* of computation is based on *stored program computer* concept where instruction and data are stored in the same memory. This architecture is the basis of most of the contemporary processors. The processor consists of a control unit, arithmetic logic unit (ALU), and registers in this model of computing

© Springer Nature Switzerland AG 2019
K. Erciyes, *Distributed Real-Time Systems*, Computer Communications
and Networks, https://doi.org/10.1007/978-3-030-22570-4_2

Fig. 2.1 Architecture of a processor

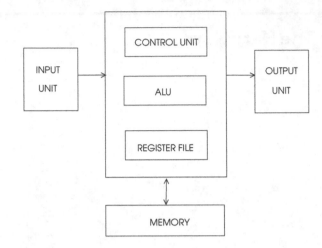

as shown in Fig. 2.1. The ALU, the register file that contains all of the registers and the interfacing circuitry is commonly called the *datapath*.

A processor first fetches an instruction from the memory and decodes it to decide on which action to perform. It then finds any data needed and finally executes the instruction and writes the produced output to the memory if needed. The *instruction set architecture* (ISA) is an abstraction to form the interface between the software and the hardware. The ISA comprises basically the assembly language of the processor and provides instructions such as *add*, *sub*, and *and* to perform operations on the processor. In other words, the ISA is the programmer's view of the hardware and specifies the memory organization, temporary memory locations within the CPU called the *registers* and the *instruction set*.

Two types of ISA-based computing are in common use: complex instruction set computing (CISC) and reduced instruction set computing (RISC). A CISC processor uses as few lines of assembly programming as possible and hence is closer to a high-level language than the RISC processor which executes simple instructions typically in one clock cycle. For example, multiplication of two numbers can be performed by a simple MULT M1, M2, M3 instruction in CISC computing which multiplies two values in memory locations M1 and M2 and stores the result in memory M3. The same multiplication needs to be broken down into several RISC instructions such as loading values from memory to two registers, adding the values in the registers inside a loop and storing the result in memory. At first glance, a RISC processor seems to be executing more lines of assembly lines than a CISC processor. Also, the compiler does less work in converting a high-level language statement to assembly language in CISC computing requiring less storage for RAM to store assembly lines. However, a CISC processor may need to work few clock cycles for an instruction whereas a RISC processor commonly finishes a line of assembly code in one clock cycle making the design of a RISC processor simpler than a CISC processor. The mode of operation of datapath can be *single cycle*, *multi-cycle*, or *pipelined* as described in the next sections.

2.2.1 Single Cycle Datapath

We will briefly describe the microprocessor without interlocked pipeline stages (MIPS) 32-bit processor datapath and its control unit as an example of a single cycle datapath. MIPS is a RISC commonly used in embedded systems and yet is simple to display the main functions of a processor. MIPS has 32 registers each with 32-bit length; hence, a 5-bit address is needed to select a register. We will first review the ISA of MIPS which has three modes: R-type, I-type, and J-type. An R-type (register type) considers the instruction word to consist of the fields shown in Fig. 2.2a.

The fields in the instruction are as follows:

- *opcode*: This 6-bit field is used to decode the instruction operation. It is the main input to the control unit that generates required control signals for the instruction.
- *rs*: 5-bit address of the first source register.
- *rt*: 5-bit address of the second source register.
- *rd*: 5-bit address of the destination register.
- *shamt*: 5-bit shift amount.
- *funct*: 6-bit function code. This field is used to perform a subfunction such as addition of an R-type instruction. It can be used as an input to the ALU.

For example, the instruction

        add $7, $14, $10

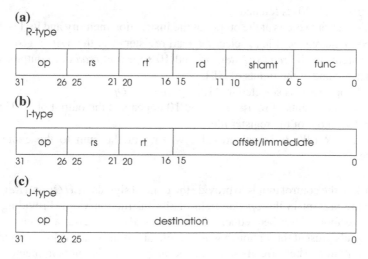

Fig. 2.2 MIPS instruction modes

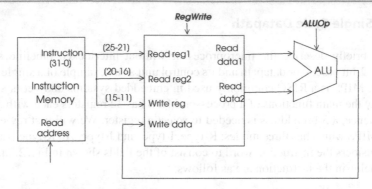

Fig. 2.3 MIPS datapath for R-type instruction

in binary and hexadecimal notation would be

```
    op         rs        rt        rd       shamt      func
  _____     _____    _____    _____    _____    _____
  00 0000   0 1110    0 1010    0 0111    0 0000    10 0000
```

The 32-bit instruction would then be 0 x 01CA:3820 in hexadecimal notation. Let us take a closer look at how this instruction is executed in the simple datapath of Fig. 2.3.

1. The read address is provided to the instruction memory and the 32-bit instruction word (01CA:3820) is fetched.
2. The instruction word is at the output of the instruction memory and its 11–25 bits are fed to the register file to select the read registers and the *write register*.
3. The contents of the *read registers* 14 and 10 appear at *read data* outputs of the register file and at the inputs of ALU.
4. The *add* operation is selected in ALU by the *ALU Op*.
5. The sum of contents of registers 14 and 10 appears at the output of ALU and at *Write data* input of the register file.
6. Activating *RegWrite* signal provides writing of the sum to the contents of register 7.

The task of the control unit is to provide the control signals *ALU Op* and *RegWrite* by the value present in the *opcode* field of the instruction word. Providing these signals in the above-described sequence is difficult and not necessary. We can have the control signals present throughout a whole cycle and if the cycle is long enough, the produced output will be correctly settled in the *single cycle datapath* implementation. Determining the length of the cycle is crucial in this design method and it should be as long as the longest instruction.

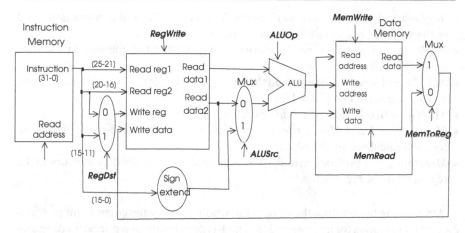

Fig. 2.4 MIPS datapath with data memory

We need to write the generated output to external data memory instead of a destination register commonly. The I-type of addressing with the format shown in Fig. 2.2 is used for this purpose and MIPS has two instructions to store data to and remove data from data memory as follows:

```
lw   $2, 12($1)
sw   $2, 12($1)
```

In this example, the *load word* (lw) instruction loads a 32-bit word from the data memory location at the address pointed by the contents of register $1+12 into register $2 and the *store word* (sw) does the opposite by storing the contents of the register $2 at data memory address $1+12. We will now extend the datapath to include I-type of instructions as shown in Fig. 2.4. There are three additional 2:1 multiplexers to be able to distinguish the instruction modes. In I-type instruction, the second register field contains the write address now, so we need to pick bits (20–16) as this value for this mode by activating *RegDest* control of the first multiplexer. There is a *sign-extend* circuit to extend the 16-bit value in the lower half of the instruction word for addition of the first register value to the contents of the address contained in the register plus the offset value. The second multiplexer selects the R-type or I-type operation by its *ALU Src* input. Lastly, the third multiplexer selects whether the contents of the data memory location (I-type) or the results of the ALU operation (R-type) are to be written to the register file by the *MemToReg* control signal or not. Loading from or writing to data memory is selected by *MemWrite* or *MemRead* signals.

Let us list the control signals needed for the *lw* 2, 12($1) instruction. We only need to supply the address of the instruction to the instruction memory without the need for a read signal since all write operations are controlled by the control unit. The opcode for this instruction which is 100011 is fed to the control unit which outputs the following signal values:

- *RegDest* = 0 since we want to interpret the bits (20–16) of the instruction word as the write reg address as this is an I-type instruction.
- *RegWrite* = 1 since we need to write to register file the contents of the data memory location.
- *ALUSrc* = 1 since we want to add the sign extended bits (15–0) of the instruction word to the value in register 1.
- *ALUOp* = 010 in binary for *add* operation.
- *MemRead* = 1 and *MemWrite* = 0 to allow reading the value contained in the location calculated by the ALU.
- *MemToReg* = 1 provides writing of the output value from the data memory to the register 2 in the register file.

Maintaining these values throughout the whole cycle by the control unit provides the correct operation for this instruction. The J-type of instruction is used for diverting from the normal flow of the program to a new instruction address. The *branch* instruction is basically used to translate an *if* statement in a high-level language and has the following format:

```
beq  $1, $2, offset
```

The values in registers 1 and 2 are compared by subtracting their values in the ALU and the address calculated by the instruction pointer (IP) value + 4 + (4 × offset) is loaded to IP if the result is 0. If these values are not equal or the instruction is not a *branch*, the value of IP is incremented by 4 to read the next location from the instruction memory. We now have two adders to be able to proceed in parallel with the ALU comparison of two input registers 1 and 2; one to calculate IP+4, and another one to calculate the branch address as shown in Fig. 2.5. Note that branch address is calculated in any case but we simply do not select it by not activating *PCSrc* signal if the branch comparison fails or we have an R-type instruction.

Control Unit

The control unit has 6-bit opcode, 6-bit function field of the instruction, and 1-bit zero output of the ALU as its inputs and it has to generate all required signals for each instruction. It can be realized by a simple combinational circuit that has the 13-bit input that provides the required control signals. Control signals for some example MIPS instructions are shown in Table 2.1.

Performance

The *load word* instruction is the longest time taking instruction in MIPS as it reads from instruction memory, then reads from register file, performs ALU *add* operation, reads from data memory, and then writes this data to register file. Each of these except the ALU operation are read and write operations resulting in memory delays. The ALU operation also takes time and the sum of these delays gives an idea on the length of the clock cycle needed. For example, assuming each of the five steps of the load word instruction takes 2 ns, and the multiplexers with zero delays, we have a clock

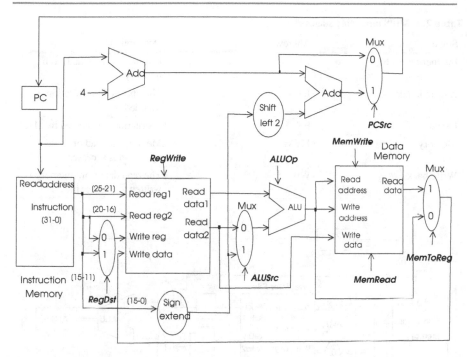

Fig. 2.5 Final MIPS single cycle datapath

Table 2.1 MIPS control signals for sample instructions

Operation	RegDest	RegWrite	ALUSrc	ALUOp	MemWrite	MemRead	MemToReg
add	1	1	0	010	0	0	0
and	1	1	0	000	0	0	0
lw	0	1	1	010	0	1	1
sw	X	0	1	010	1	0	X
beq	X	0	0	110	0	0	X

with 10 ns cycle time and 100 MHz frequency. However, many MIPS instructions require less time, for example, R-type instructions do not use data memory and hence execute in four steps requiring a clock cycle of 8 ns. In reality, accessing external data memory is much slower and hence the single cycle datapath has poor performance.

2.2.2 Multi-cycle Datapath

Single cycle datapath is not efficient as it uses the same length clock for all instructions. Our aim in improving the single cycle implementation is to have variable length

Table 2.2 MIPS pipelining stages

Stage	Abbrev.	Description
Instruction fetch	IF	Instruction is fetched from memory
Instruction decode	ID	The opcode in instruction is decoded
Execute	EX	Instruction is executed in ALU
Memory	MEM	Memory for read or write operation is accessed
Write back	WB	The read data from memory is written to register file

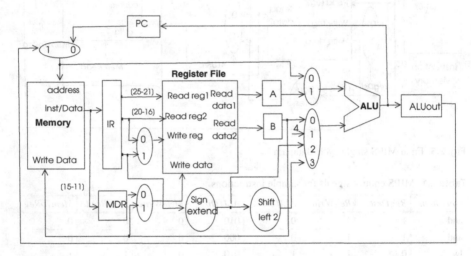

Fig. 2.6 MIPS multi-cycle datapath without the control signals

of execution time for each instruction. We will start by dividing the instruction in MIPS into the stages of Table 2.2 which each stage taking one clock cycle.

The instructions now take multiple cycles, branches and jumps take three cycles, register instructions take four cycles, and only the *load word* instruction uses all five steps above. Some components of the datapath need to be reused necessitating rearrangement of the hardware. Extra registers are needed now to hold intermediate results between cycles. Instruction and data memory can be merged and we can have a single ALU that will also handle branch and jump instructions simply because the processor cannot use these hardware components simultaneously as these are employed at different cycles. The new multi-cycle datapath without control signals is depicted in Fig. 2.6. The newly introduced registers are the instruction register (IR), memory data register (MDR), two registers A and B at the output of the register file, and the ALUout register to hold the value obtained from the ALU to write back to data memory.

The control unit of a multi-cycle may be realized using a finite state machine (FSM) which consists of a finite number of states and transitions between the states. Informally, an FSM receives inputs in a state and based on its current state and the inputs, it produces output and may change state. We can have each multi-cycle stage represented by a state of the FSM and the control signals needed in that stage will be output by the FSM [1]. Although multi-cycle datapath results in a better performance than the single cycle one, the pipelining described next improves performance better and is the basis of modern processors.

2.2.3 Pipelining

Pipelining is commonly employed in real life to improve performance. Consider a washing system that has three machines; a washer (W), a drier (D), and a folder (F). We will assume a person needs to use these machines in sequence as washer, drier, and folder. Let us further assume the washer, the drier, and folder take 20 min each. Allocating all of these machines to a person X as shown in Fig. 2.7a is not sensible since the washer, for example, can be used by another person Y at the drying stage of person X. Pipelining shown in Fig. 2.7b allows the unused machines to be used by the next person and reduces the completion of the laundry process for three persons

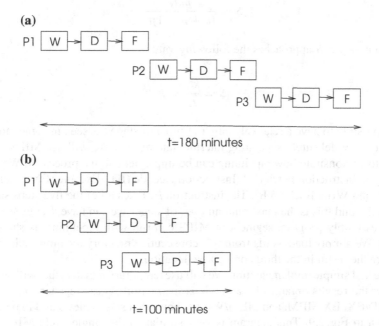

Fig. 2.7 Pipelining example. Three machines in a laundrette are used without pipelining in (a) to result in a total time of 180 min. Pipelining reduces the completion time to 100 min as shown in (b)

to 100 min. Note that pipelining does not reduce the time for each person as we still have 60 min of processing each, but it reduces the overall completion times for tasks.

One important finding of the above example is that we need to clearly specify parts of the task to be performed in sequence to be able to employ pipelining. These stages were washing, drying, and folding for the laundry of each person. In computational terms, the instruction is to be executed in a pipelined processor and we need to separate the execution into a number of distinct phases. The theoretical speedup obtained by a pipelined processor can be determined as follows:

- Let k be the total number of pipeline stages, t_s is the time of processing at each stage, and n is the total number of tasks presented to the pipeline.
- The first instruction takes kt_s time to complete.
- We have an instruction coming out of the pipeline at each cycle for the remaining $(n-1)$ tasks resulting in a total $(n-1)t_s$ time to finish the remaining tasks.
- Total time for $(n-1)$ tasks is then

$$kt_s + (n-1)t_s = (k+n-1)t_s$$

- The speedup S obtained is the ratio of the sequential time without a pipeline to the pipeline processing time of n instructions as follows:

$$S = \frac{nkt_s}{(k+n-1)t_s}$$

When $n \to \infty$, S approaches the following value:

$$S = \frac{kt_s}{t_s} = k$$

It is possible to have a relatively larger number of stages as seen in some modern processors, by delicately separating intermediate results. We will use MIPS architecture to demonstrate how pipelining can be implemented in a processor. We have five stages: Instruction Fetch (IF), Instruction Decode (ID), Execute (EX), Memory (MEM), and Write Back (WB). The instruction lw takes all of the five steps shown in Table 2.2 and this is the maximum number of stages we can have. Let us see how a short assembly program segment in MIPS can run with pipelining as shown in Fig. 2.8. We simply load words from two consecutive memory locations, add them, and store the result in the third consecutive location.

In the first simple implementation, we will use temporary registers that will be used to buffer the results obtained by a previous instruction. These registers are named IF/ID, ID/EX, EX/MEM, and MEM/WB to show what stages they stand in between as shown in Fig. 2.9. This diagram is very similar to the single cycle MIPS with minor modifications. For example, calculation of the *branch* (beq) instruction can be performed in parallel with the ALU computation, and calculating the next address is performed while fetching the instruction since we have the address information in

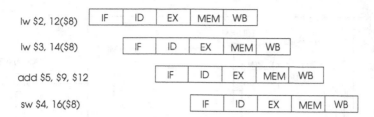

lw $2, 12($8)

lw $3, 14($8)

add $5, $9, $12

sw $4, 16($8)

Fig. 2.8 Pipelining in MIPS assembly language program

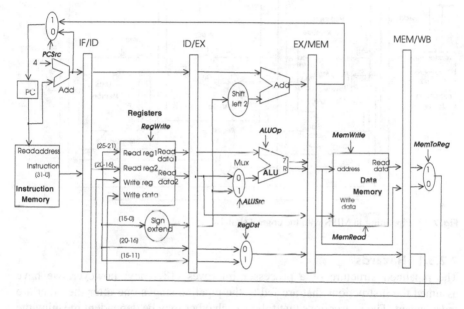

Fig. 2.9 Pipelining in MIPS

the first stage. A data value required in a later stage must be forwarded between the pipeline registers. For example, the write register address is carried all through the stages since this is needed in the last stage in the case of *store word* (sw) instruction. Note that an instruction goes through one stage at each cycle requiring five cycles for a whole instruction.

Control Signals

Control signals needed for the instruction should be transferred along with the instruction using the pipeline registers. There are distinct control signals needed at each stage and once the instruction passes through a stage, control signals needed for that stage can be abandoned resulting in fewer signals to be propagated as the instruction goes through the stages as shown in Fig. 2.10.

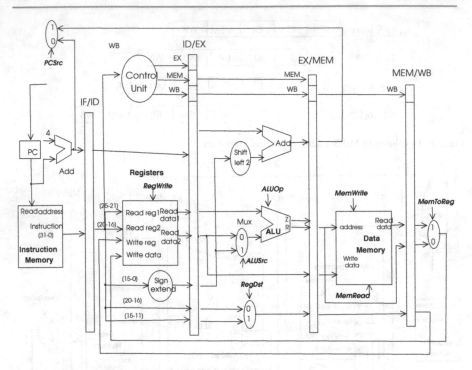

Fig. 2.10 Pipelining in MIPS with the control signals

2.2.3.1 Hazards

The pipelined structure of a processor improves efficiency; however, we have assumed the instructions that are fed to the pipeline stages one after the other are independent. The instructions that follow each other may be dependent meaning the value of a register produced is used in one of the following instructions which cause hazards.

Data Hazards

Consider the following lines of MIPS code:

```
lw    $3, 10($2)   First example of a data hazard
        \
and   $4, $3, $5

add   $3, $1, $2   Second example of a data hazard
        |
sw    $3, 12($4)
```

The first instruction load word (*lw*) loads the value in data location ($2)+10 into register 3 which is done at the final and fifth write back (WB) stage. The following instruction however needs the value in register 3 in the third execute (EX) cycle.

This situation is termed *data hazard* in a pipelined processor. A possible solution is forwarding the value of $3 which is fetched from memory in the fourth memory (MEM) stage of the *load word* instruction to the *and* instruction that needs this value in its second instruction decode (ID) stage. This type of forwarding results in handing over of the register value to two stages back. In order to accomplish correct operation, we need to stall the processor for one cycle between these two instructions. Stalling a processor means stopping its operation for one cycle by resetting the values in the required pipeline registers.

In the second example above, the store word (*sw*) instruction needs the correct value of register 3 at the fourth MEM stage, and hence this value can be forwarded by the add instruction at its second EX stage without stalling the processor. When and where to forward register values can be handled by a *forwarding unit* which is a circuit which compares the destination register (*rd*) value at the end of the EX stage (EX/MEM pipeline register) and also at the end of the MEM stage (MEM/WB pipeline register) with the values of the first (*rs*) and second (*rt*) register values at the end of the ID stage (IF/ID pipeline register). Whenever a match is found the matched register value is forwarded one or two stages back depending on where the match is found. The comparisons to be performed at the forwarding unit are as follows. A detailed description of the forwarding operation in MIPS is given in [1].

- EX/MEM.rd = ID/EX.rs
- EX/MEM.rd = ID/EX.rt
- MEM/WB.rd = ID/EX.rs
- MEM/WB.rd = ID/EX.rt

The MIPS datapath with the forwarding unit is depicted in Fig. 2.11. This unit basically compares the destination register (*rd*) identifier input to MEM stage with the input registers of the EX stage (*rs* and *rt*), and the destination register (*rd*) identifier input to WB stage with the input registers of the EX stage (*rs* and *rt*). If a match is found, the *rd* value is fetched back one or two stages. This carrying back of the *rd* value is accomplished by selecting the 1 (2 stages back) or 2 (1 stage back) inputs of the multiplexers now present at the input of the ALU.

Control Hazards

The sequence of instructions may divert to another location after the branch (beq) instruction which compares values in two registers and jumps to the location specified in the instruction if these values are the same. Instructions are fetched in the first stage (IF) and the branch decision is made at the third stage (EX). Consider the following MIPS code segment:

```
beq   $1, $2, 16($6)

and   $4, $3, $5

add   $7, $8, $9
```

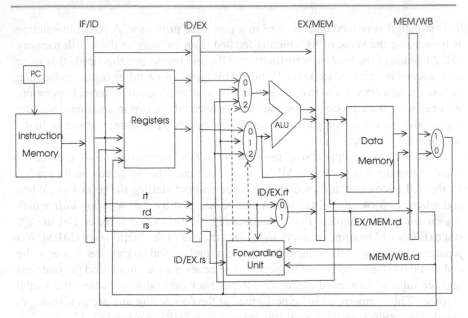

Fig. 2.11 MIPS forwarding unit

The branch may or may not be taken depending on the values in registers 1 and 2 and this is decided in the EX stage by the ALU operation which would subtract the values in these registers and the zero flag being set at the end of this operation means the instruction address should be changed. If the branch is taken, the *and* and *add* instructions which are at stages ID and IF stages, respectively, should be abandoned. Flushing of these stages by resetting the IF/ID and ID/EX pipeline registers is one solution to this problem at the expense of lost CPU cycles. Main solutions to the control hazards for pipelined processors are as follows:

- *Stalling the Processor*: Stop the pipeline until the outcome of the branch operation is known.
- *Branch Delay Slots*: The few instructions that follow the branch instruction are always taken. The compiler or the programmer puts useful instructions or no operations (NOPs) in these slots. For processors with a large number of pipelines, this solution is not convenient.
- *Branch Prediction*: The outcome of the branch instruction is predicted using some heuristics. *Static branch prediction* can be implemented by always taking the branch or not. An improvement is achieved by the *backward taken/forward not taken* principle assuming backward offsets are likely loop jumps that will be taken most of the time and forward jumps are rarely taken. Dynamic branch prediction uses current status to decide to take the branch or not.

2.2.4 Microcontrollers

A *microcontroller* has the processor, memory, and I/O interface in a single chip. Microcontrollers are commonly used for low-power and small-sized embedded applications. A *field-programmable gate array* (FPGA) is a type of dynamically reconfigurable microcontroller. The application developer can construct the hardware required for the real-time system using the FPGA. These processors are commonly used in communication systems and image processing. *Digital signal processors* (DSPs) are special purpose microprocessors used for signal processing which requires frequent arithmetic operations on large amounts of data.

2.3 Memory

Memory, along with input/output is the main bottleneck that bounds the speed of data transfer in a computer. There is always a need for large and also fast storage and access of data. Large storage is needed in contemporary computers because data is available in unprecedented large amounts now and the size of available data increases in daily basis. On the other hand, if access to memory is not fast enough, we cannot exploit the full capabilities of a high-speed processor.

A static random access memory (SRAM) is fast and expensive whereas a dynamic RAM (DRAM) of the same size is more than a magnitude cheaper than SRAM but much slower. An SRAM 1-bit cell consists of two gates, whereas a 1-bit value is stored as a charge on a capacitor in a DRAM. An erasable programmable memory (EPROM) retains program code and data when power is switched off. A hard disk can hold data in the order of terabytes but access to it is very slow although it is the cheapest in memory hierarchy. We need ways to incorporate efficient, fast memory access.

2.3.1 Interface to the Processor

The communication medium between the processor and the memory consists of three buses; the address bus to locate the address of data, data bus to transfer data, and control bus for control operations. Typical control signals for a RAM memory are read (RD) and write (WR) lines which enable read and write operations from and to the memory, respectively, as displayed in Fig. 2.12. EPROM needs only RD signal since it cannot be written during normal operation. For n-bit address and m-bit data lines, the processor can address 2^n memory locations each of which is m-bits long. The high-order bits of the address bus is commonly used to select the memory and the device that is interfaced to the bus. For example, a 10-bit address bus (A_9-A_0) and an 8-bit data bus allow 1024 memory locations to be addressed. Consider dividing this space into 512 bytes of EPROM and 512 bytes of RAM, we can use the bit A_9 as the chip select input to the memories. In this case, data in the address range

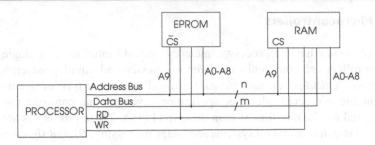

Fig. 2.12 Memory interface to the processor

0–511 is in EPROM and data in 512–1023 addresses resides in the RAM. Note that the chip select input of the EPROM is active low $(\overline{CS})$ to enable it when $A_9 = 0$. Contemporary processors are commonly interfaced in hierarchical bus structures to the memory blocks and to the I/O interfaces.

2.3.2 Caches

A cache is a static RAM of small size, placed between the processor and the dynamic RAM to provide fast access to instructions or data. The idea of the cash is to keep the most frequently used data there so that retrieving it from a closer location is faster than transferring it from the DRAM. The principle of *temporal locality* assumes an accessed address will be accessed by the processor again with high probability in near future as in the case of loops. On the other hand, the *spatial locality* principle assumes that the neighbor addresses of already accessed locations have a high chance of being accessed which is a valid argument since a program executes sequentially most of the time. Note that a loop also has spatial locality since the loop line sequence is executed repeatedly until the loop terminates. A *cache hit* is the situation when the cash contains the data we are searching and a *cache miss* occurs otherwise. Cache performance is measured by the *hit rate* which is the percentage of cache hits over total memory accesses. The time taken for data to be sent from cache to the processor is the *hit time* and the time needed to transfer data from the main memory to the cache when miss occurs is the *miss penalty*. The percentage of misses is the *miss rate* and based on these parameters, the *average memory access time* for a computer with a cache is equal to the hit time + (miss rate × miss penalty).

Spatial locality principle can be exploited by copying a block of data that contains the neighbor locations as well as the required location to the cache. With this principle, data is read and written to a cache in blocks each having an *index*. Assuming a cache has 2^k blocks, we need a k-bit long index to address a block. Whenever a memory location *addr* is addressed, the lower k bits of *addr* can be used to find the corresponding block in cash. Moreover, we need to check whether the searched address corresponds exactly to the cash address. A block contains a *tag* value to match the remaining address bits; hence, if the tag value of the block equals to the

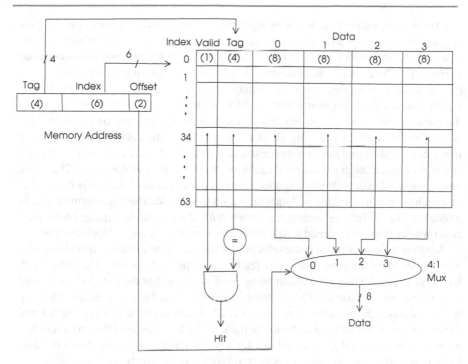

Fig. 2.13 Cache interface

remaining higher order bits of the address, we have a hit. We also have a *valid bit* for each cache block to indicate there is valid data in the block or it is empty. For example, let us consider a main memory of 12-bit addresses which is divided into 4-byte blocks giving a total of 4 KB memory. Each block can be addressed by 6 bits; therefore, a cash with 6-bit index value and a 4-bit tag value is needed as shown in Fig. 2.13. Since each block has 4 bytes, the lower 2 bits of the memory address is used as the offset within the block to access the correct byte.

The operation of the cache is shown in pseudocode in Algorithm 2.1.

Algorithm 2.1 *Cache*

1: Upon generation of a memory address *addr* by the processor
2: **if** *valid* bit of the block pointed by the *addr* index bits = 1 **then**
3:    **if** *tag* of *addr* = *tag* of cache block **then**
4:       **output data** from cache at base location of the block + *offset* of the memory address
5:    **else get data** from external memory and hand it to the processor
6:       write data to cache by removing an unused entry
7:    **end if**
8: **else get data** from external memory and hand it to the processor
9:    write data to cache in the empty location
10: **end if**

In reality, we do not have to run an algorithm for every memory access as this process can be handled by an electronic circuit called the *cache controller* which consists of an k-bit comparator for a k-bit tag, an *and* gate, and a 4:1 *multiplexer* as depicted in Fig. 2.13. The cache we have described up to now is called a *direct-mapped cache* in which every main memory location maps to exactly one cache block location of which can be calculated from the physical memory address. Data transferred from the memory can be written to any unused location in a *fully associative cache*. This type of cache requires checking all of the locations of the cache in parallel using many comparators and hence is expensive to construct. A *set-associative cache* is a solution between the direct-mapped cache and the fully associative cache. The cache now consists of groups of blocks called *sets* and each memory location is mapped to one such set but the location of data within a set is variable. The placement of data in a block is like the fully associative cache allowing data search in smaller cache area. A set-associative set is named k-*way associative cache* if it has k blocks in a set.

Another issue of concern is the selection of the data to be removed from the cache when a new data from the memory is fetched and the cache is full. The data which has stayed in cache longest without being used is selected as the data to be removed in the *least recently used* (LRU) method. A doubly linked list may be used to keep track of data usage in cache. A newly referenced data is put in front of the list and the data to be removed is taken from the back of the list. Various other policies such as *first-in-first-out* and *last-in-first-out* can also be employed to decide on the data to be removed as the first and last accessed blocks, respectively.

When a processor writes data to cache, it must also write the data to the main memory for consistency. The *write policy* determines how this procedure is performed as follows:

- *Write Through*: The cache and the storage are written synchronously.
- *Write Back*: The cache is written first and writing to memory is delayed until the modified data in cache is replaced by another cache block.

2.4 Input/Output Access

The processor communicates with the external devices using input and output *ports* which are basically data buffers to interface to the processor data bus. Input/Output (I/O) devices come in many forms, from keyboards to printers. I/O devices are many orders of magnitude slower compared to the processor and to memory. Disks and networks are the two main I/O systems that communicate frequently with the processor. In order to increase the speed of data transfers from/to I/O devices, parallelism in hardware can be employed. For example, redundant array of inexpensive disk (RAID) systems provides parallel access to a number of hard disks at the same time, and memory can be divided into banks that are accessed in parallel. We will first review common interface to I/O devices in real-time systems and then describe methods of I/O interface in hardware and software in this section.

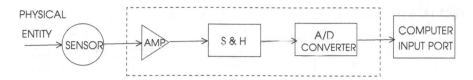

Fig. 2.14 Input interface of a real-time system

2.4.1 The Input Device Interface

Two types of I/O devices are prevalent in real-time systems: *sensors* and *actuators*. A *sensor* is an input device capable of sensing an external physical parameter and producing an electric signal representing this parameter. There are many sensor types such as temperature, humidity, acceleration, and pressure sensors. Typical parameters for a sensor are its operating temperature, error specifications, and range of the values it can sense.

The *input interface* to the processor contains electrical circuitry to convert the signal from a sensor to a form suitable to be processed by the real-time computer. The output from the sensors is commonly small in magnitude in the range of millivolts needing analog signal amplification and then employing an analog-to-digital (A/D) converter which transforms the electrical signal to digital data. Sensors may have the needed converters embedded in them. The structure of a typical input interface employing A/D converter is displayed in Fig. 2.14 where an *amplifier* (AMP) magnifies the voltage level obtained from a sensor which would typically be in millivolts range, and a *sample and hold* (S & H) circuit provides discrete instants of the input signal to the A/D converter so that the converter has a stable input during conversion. The number of bits output by the A/D converter shows the precision of the input signal. A 32-bit A/D converter produces digital data in the range of 0–65535 for a typical 0–5 volt analog input range.

A detailed typical interface of an A/D converter to the processor is depicted in Fig. 2.15. The processor initiates conversion by the *start* pulse when it needs to input sensor data. This pulse may be active high or active low, or sometimes just a transition between high and low levels depending on the A/D device characteristics. Generating a pulse by the processor is simply performed by writing a 0, and then 1 and finally 0 to *start* bit by the output port. A delay after each write is needed to form a pulse. The width of the pulse must be greater than the minimum pulse width needed by the A/D converter and this parameter can be adjusted by the length of the delay, which is commonly done by loading a register with a value and decrementing it until it reaches 0. The output from the converter is available when *end-of-conversion* (EOC) bit is set. This signal can be connected to the input port of the processor which can be checked by polling this line continuously or it may be used to interrupt the processor.

Fig. 2.15 A/D interface

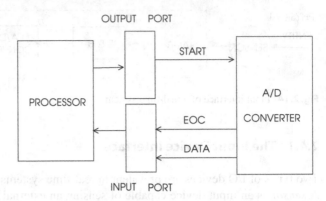

2.4.2 The Output Device Interface

An *actuator* works in opposite way to a sensor by converting the electric signal to some form of physical parameter such as sound, heat, or motion. For example, a solenoid is an actuator which produces a magnetic field when current runs through it. In order to start an actuator, the binary output data is first converted to an analog signal by the use of a digital-to-analog (D/A) converter. This signal may be transferred to a convenient form by a signal conditioning circuit to activate an actuator as shown in Fig. 2.16. The signal conditioning may include an amplifier and circuitry to limit the value of the analog signal. Its operating temperature and the range of outputs it can produce are the main characteristics of an actuator.

2.4.3 Memory-Mapped and Isolated I/O

Communication between the processor and the I/O peripherals can be performed using two fundamental methods: *Memory-mapped I/O* and *isolated I/O*. In memory-mapped I/O, the address space is divided into memory locations and I/O peripherals. An I/O unit is accessed just like a physical memory address of memory with I/O units connected to a common bus. The processor sends address of the I/O unit and data over the common bus and an I/O peripheral listening over the bus reacts when

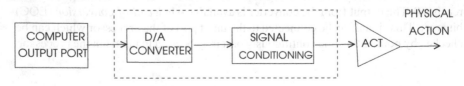

Fig. 2.16 Output interface of a real-time system

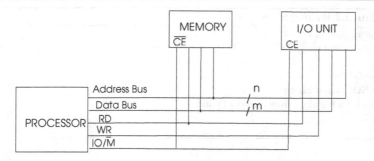

Fig. 2.17 Isolated I/O interface to the processor

it is addressed as shown in Fig. 2.17. The isolated I/O approach provides distinct I/O instructions that may use the same address space as the memory. In this case, the processor should supply a separate control signal that shows whether the intended operation is a memory or an I/O operation. Isolated I/O can be convenient when there are numerous I/O peripherals to be addressed since the same address space can be used for I/O operations. On the other hand, memory-mapped I/O is flexible since all of the instructions used for memory operations can be used for I/O operations. A typical control signal used in isolated I/O is the $IO/\overline{M}$ signal by Intel line of processors which shows I/O instruction when activated and memory, otherwise, as shown in Fig. 2.17.

2.4.4 Software Interface to I/O

Having discussed how I/O peripherals can be interfaced to the processor in hardware, we can now review basic software methods to access the I/O units. The processor reads some data from an input device and writes to an output device which are slow. Two main methods of software interface to I/O devices are the polling and interrupt-driven I/O methods.

2.4.4.1 Polling

Polling is a basic software interface method to an I/O device in which the processor makes a request to the device and checks continuously whether the request is fulfilled. A typical writing of a file by the processor to the hard drive using this method is shown in Algorithm 2.2. The CPU sends a *write* request to the disk and the disk raises a control signal when it is ready with the disk head in the right sector, etc. This signal is continuously checked by the processor and when found active the transfer of the file is initiated. An obvious disadvantage of this method is the wasted CPU cycles while waiting for a slow device.

Algorithm 2.2 *Writing to Disk*

1: **send** *write* request to disk
2: **while** disk is not ready **do**
3:     *wait*
4: **end while**
5: **while** NOT end of file **do**
6:     **transfer** a file block from memory to disk
7: **end while**

2.4.4.2 Interrupt-Driven I/O

An *interrupt* is initiated by an external source and the processor halts its current processing when this happens. It serves the interrupt and then continues with the task it has stopped processing. This type of operation is analogous to an everyday example in which the doorbell rings while a person in the house is reading a book. The person would place a marker in the book (save current environment), respond to the visitor (serve the interrupt), and then continue with reading. The handling of an interrupt by the processor is similar to the described example; the current running environment which consists of register values, data, file pointers, etc. are saved; the action requested by the interrupt is performed, and then the environment is restored and processing is continued from the position (address) it was stopped. All of this processing is commonly handled by the operating system.

Returning to the disk example, the processor makes the *write* request as before but continues with its other processing until an interrupt by the disk showing it ready is received. The interrupt is then serviced by writing to the disk and when this process is over, the stopped task is continued. An interrupt service routine may or may not be interrupted. All steps of this operation is under the control of the real-time operating system and it is possible to have variations such as having a multi-layered interrupt service routine (ISR) with a high and a low priority sections of code in which the high priority part is served first.

A processor typically has an interrupt activation pin (commonly labeled as *INT* or $\overline{INT}$) which may be activated by an external device. This line is checked during every instruction of the processor and when found active, a service routine at a predefined location is activated. Generally, a processor has more than one interrupt pin one of which is designated as the *non-maskable interrupt* and another that can be disabled by the software called *maskable interrupt*. Serving an interrupt can be achieved using different approaches. In the *interrupt polling* method, all of the device interrupt requests are wire-ORed to the processor INT input and when this line is activated, the processor may either read status register of each device over a common bus or more commonly can check an input port to find the source of the interrupt as shown in Fig. 2.18.

We have three devices connected to the processor and a device requesting service will activate its REQ line and place a 1 on the input port line of the processor it is connected. The processor detects the active port line upon the INT line becoming active as shown in Algorithm 2.3 and goes to the memory location where the ISR for

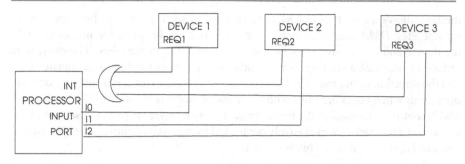

Fig. 2.18 Interrupt polling

that device is kept. If there are more than one interrupts occurring simultaneously, a priority method may be used.

Algorithm 2.3 *Polling ISR*

1:
2: upon activation of INT
3: **input** *port1* to *reg*1
4: *count* ← 0
5: **load** 1 to *reg*2
6: **while** *count* < 3 **do**
7:     **and** *reg*1 with *reg*2, store result in *reg*3
8:     **if** *reg*3 ≠ 0 **then**
9:         goto ISR location *base* + *count*
10:         **exit**
11:     **end if**
12:     **shift** *reg*2 left once
13:     *count* ← *count* + 1
14: **end while**
15:

A different and a faster approach is the *vector interrupt* method in which the interrupt source activates the INT input and places the address of its ISR on the input port or data bus of the processor. Note that the task of serving the interrupt is simpler in this case by simply loading the new instruction address from the bus contents by the processor. In a more practical approach, a device requesting service puts its offset from a common interrupt table base resulting in a simpler interface to the processor. For example, 5 input bits will be adequate for 32 devices.

2.4.4.3 Direct Memory Access

Devices such as disks and networks operate at faster rates than devices like keyboards and they require transfer of large blocks of data. Direct memory access (DMA) method allows transfer of large blocks of data to such devices at the speed they operate. The *DMA controller* is an interface processor between the processor and the device and communicates with the processor for data transfers. Typically, the processor provides a starting memory address, number of bytes to be transferred, and the direction of the transfer. The processor will have to refrain from any external operations when buses are busy with the transfer. In *cycle stealing* DMA mode, the DMA controller takes over the buses for each byte transfer when the buses are idle. Conversely, the buses are exclusively controlled by the DMA controller until the data transfer is completed in the *burst* mode.

2.4.4.4 Exceptions

Exceptions are internal interrupts which are detected by the processor. Illegal instructions, attempting to divide by 0 and arithmetic overflow are the common causes of exceptions. In the most basic form, the program that caused the exception will be stopped and removed from memory by the operating system; however, there are cases when recovery is possible. Commonly, the operating system takes over when an exception is detected with the processor notifying the cause of exception and the instruction that caused it.

The MIPS processor has the exception program counter (EPC) which contains the address of the instruction that was being executed when exception occurred. The *Cause* register has bits that help to identify the cause of the exception. When an interrupt or an exception happens in MIPS, the following are performed:

1. Current program counter is transferred to EPC.
2. The cause of exception is stored in *Cause* register.
3. Exceptions and interrupts are disabled by modifying the status register.
4. Branch to the exception/interrupt handler address is taken.

Upon returning from the handler, the contents of the EPC register is moved to the program counter and interrupts and exceptions are enabled by modifying the status register.

2.4.4.5 Timers

A *timer* is commonly used in a real-time system to measure a time interval. It is loaded with a value corresponding to the interval and at a certain number of clock ticks, the counter is decremented. When the value of the timer reaches zero, an interrupt to the processor is generated to indicate that time has elapsed. A timer can also be configured to count from 0 value upward and when its value reaches the value for the interval, an overflow bit generated by the timer can be used to activate the interrupt line of the processor.

This type of *timer interrupt* is convenient in a real-time system in which the main concern is to have the tasks meet their deadlines. The timers in embedded systems are commonly called *watchdog timers* and are used to reset or start a recovery process when there is a fault in the system. The processor regularly resets the count of the watchdog timer in shorter intervals than the time it takes for the watchdog timer to overflow. Hence, there is a hardware or software error if this timer overflows which can be used to interrupt the processor to start a recovery process. The timer interrupt mechanism also finds many applications in non-real-time systems.

2.5 Multicore Processors

A *multicore processor* has two or more processor embedded in the same chip to enhance performance. Many contemporary processors are multicore. A multicore processor may have homogeneous/symmetric core processors as in the case of a general central processing unit, or it may have heterogenous/asymmetric cores as in a graphic processor. The cores employ a hierarchy of caches to improve performance; L1 and L2 caches are typically used by a core and L3 cache is shared between the cores as shown in Fig. 2.19 where a sample quad-core processor is depicted with L1 caches assigned to instruction and data.

A multicore processor improves performance mainly by concurrent execution of tasks in the cores. Power used is reduced compared to two or more processors in separate chips which prolongs battery life. However, sharing of resources such as caches and buses requires protection from concurrent access. Moreover, tasks commonly called *threads* running on cores need to be synchronized, similar to task synchronization in a general operating system running on a one-core processor.

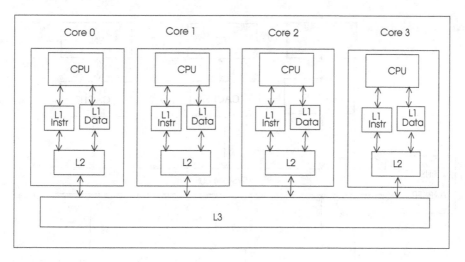

Fig. 2.19 A quad-core processor

2.6 Multiprocessors

A *multiprocessor* has two or more processors that share memory and I/O devices. A *symmetric multiprocessor* (SMP) system has homogenous processors that share a global memory. An SMP system is also called a *tightly coupled multiprocessor* system since there are frequent data transfers between the processors using the shared memory, in contrast to a distributed system where the main transfer method is via the exchange of messages. An SMP system with *n* processors and *k* I/O devices interfaced to a common bus is shown in Fig. 2.20 with a *bus arbiter* that controls the use of buses.

Multiprocessor computing systems can be classified into two main branches as follows:

- *Single-Instruction-Multiple-Data* (SIMD) systems: A single control unit controls a number of datapaths in this model. The same instruction is executed with many data elements in parallel in SIMD computers to achieve high performance. SIMD computers have special hardware and are used for scientific computations requiring high volumes of data which frequently involve matrix operations.
- *Multiple-Instruction-Multiple-Data* (MIMD) systems: These multiprocessors share data in *centralized memory* systems and communicate using messages in *distributed memory* multiprocessor model.

When the communication medium in a distributed memory multiprocessor system is a computer network enabling computing nodes to be physically apart, we

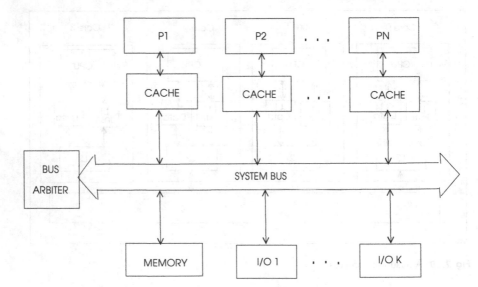

Fig. 2.20 An SMP with *n* processors, P1, …, PN and *k* I/O devices, I/O 1, …, I/O K

have a distributed system and the word multiprocessor is commonly not mentioned in this case. A centralized memory multiprocessor system usually has a single operating system that manages the basic management tasks over the system. These basic functions are the scheduling and management of tasks in multiprocessors, inter-task synchronization, and communication and protection of common resources such as the global memory during concurrent accesses.

2.7 Review Questions

1. Name the essential components of a processor.
2. What is the instruction set architecture (ISA) and what is its significance?
3. If you were to design a processor, what would be your starting point?
4. What is the main idea of a single cycle datapath?
5. How would you compute the length of the clock cycle in a single cycle datapath?
6. What is the rationale of designing the multi-cycle datapath?
7. Name the basic stages of the MIPS pipelined datapath.
8. What are data hazards and control hazards in a pipelined datapath? Give an example of each in MIPS assembly language.
9. Compare memory-mapped I/O and isolated I/O in terms of assembly language programming capabilities and the number of I/O devices used in an application.
10. What is a cache?
11. Why is DMA needed?
12. What is the difference between an I/O port, an I/O interface, and an I/O device? Give an example of each.
13. What is the difference between an interrupt and an exception?
14. Compare the main software problems encountered in a multicore processor with a multiprocessor system.

2.8 Chapter Notes

We reviewed basic hardware components of a real-time computation node. The basic datapath types are single cycle, multi-cycle, and pipelined datapaths. A single cycle datapath has the same cycle duration for all instructions and hence is not efficient. A multi-cycle datapath improves performance by having each instruction taking different clock cycles. The instruction is executed in various stages in a pipelined processor. These stages are instruction fetch (IF), instruction decode (ID), execute (EX), memory (MEM), and write back (WB) in the MIPS processor. We based our review in the analysis of processor hardware on the MIPS processor which is commonly used in embedded system design.

Memory is another fundamental component of the hardware of a real-time node. Caches at various levels of proximity to the processor are used to improve performance. Frequently used data may be kept in a cache to access it in a less costly way in future transfers. The I/O interface to the processor is handled by the input and output ports which are used to buffer the data from and to these devices. Common I/O devices connected to a computer node are the display units, keyboard, disks, and the network interface devices. Two main methods of software interface to the I/O are polling in which a device is polled by the processor until it is ready for data transfer, or use of interrupts. An interrupt is produced by an external device by activating a signal line of the processor which stops, performs the required action by an interrupt service routine, and returns to where it has stopped executing. Serving interrupts efficiently is an important issue in real-time systems as these occur frequently and meeting deadlines in the presence of interrupts is important.

A multicore processor has two or more processing units called cores in the same chip which share data through caches. Modern processors commonly employ multicore technology for performance as the limits in clock speed are being reached. A multiprocessor is a magnified view of a multicore processor in a way, in which L3 cache is replaced by a global memory and the cores are now the processors. Task synchronization and protection of global memory are the main problems to be handled by the multiprocessor operating system.

2.9 Exercises

1. Design a combinational circuit that will perform as the control unit of the single cycle datapath MIPS for the instructions *load word*, *store word*, and *add*.
2. Describe the hazard in the following MIPS code and propose a solution to remedy this hazard in MIPS.

```
sw    $1, 16($6)
and   $1, $2, $3
add   $5, $7, $6
```

3. Describe the hazard in the following MIPS code and propose a solution to remedy this hazard in MIPS.

```
beq   $3, $4, 8($12)
add   $1, $2, $7
sub   $5, $6, $8
```

4. Assume a processor with 20-bit address bus and a 16-bit data bus. Sketch the interface diagram of this processor to yield a 256 KB EPROM in [0–256) KB locations, 256 KB SRAM in [256–512) KB locations, and a 512 KB DRAM in [512–1024) KB locations.
5. Draw the circuit diagram of the interface for a vectored interrupt of three devices to a processor. Write the interrupt service routine for the processor in pseudocode.

Reference

1. JL Hennessy, DA Patterson (2011) Computer architecture: a quantitative approach, 5th edn. Morgan Kaufmann

Distributed Real-Time Systems

3

3.1 Introduction

A distributed real-time system (DRTS) consists of autonomous computing nodes
connected by a real-time network. Nodes in such a system cooperate to achieve a
common goal within specified deadlines. Distributed real-time systems are needed for
a number of reasons. First and foremost, real-time computing is naturally distributed
with nodes processing parts of an application that may be apart. Second, one of the
most important reasons in employing distributed systems is to achieve fault tolerance
which is a fundamental requirement from any real-time system. Also, balancing the
load across the nodes of a DRTS improves performance.

The computation performed at each node should meet timing constraints of the
tasks. Moreover, the network must provide real-time processing with bounded message delays. Many real-time applications are distributed in nature where tasks performed by a node at such a system are affected by tasks run at other nodes in the
network. Tasks need to communicate and synchronize in a DRTS over the real-time
networks. A modern car is equipped with such a DRTS where sensing nodes for
temperature, speed, water, oil levels, etc. are connected over a real-time network.

We start this chapter by reviewing models of DRTSs. We then have a brief survey of
the functions of a distributed real-time operating system and the main requirements
of middleware of a DRTS. Real-time communication forms the framework of a
DRTS. We describe models of real-time communication with emphasis on topologies
and data link layer protocols. We have a brief review of commonly used real-time
protocols and then conclude the chapter with two exemplary DRTS architectures.
This chapter mainly serves as an introduction to various DRTS concepts introduced
throughout the remainder of the book.

© Springer Nature Switzerland AG 2019

K. Erciyes, *Distributed Real-Time Systems*, Computer Communications
and Networks, https://doi.org/10.1007/978-3-030-22570-4_3

3.2   Models

The general architecture of a distributed real-time system is depicted in Fig. 3.1. Each node of the network is responsible for certain dedicated functions and needs to communicate with other nodes over the real-time network to perform its functions and produce the required output. A fundamental distinction in the design of a DRTS is to aim at a time- or event-triggered system.

3.2.1   Time- and Event-Triggered Distributed Systems

We will describe time- and event-triggered systems in the context of a DRTS. A *trigger* is an event that initiates a response in a real-time system. System must respond to external events at predefined instants in *time-triggered systems* as stated before. For example, measuring of temperature of a liquid at every 5 s is performed in a time-triggered system. Scheduling of the events can be performed offline as the time and duration of the execution is known beforehand with insignificant overheads at runtime. Testing and fault tolerance can be performed more simply in a time-triggered DRTS as the characteristics of the failing hardware or software module is known beforehand to replace it with an exact replica. Time-triggered systems are suitable for periodic hard real-time tasks; however, they have very little flexibility when there is a change in the task set and the load experienced by the system.

A time-triggered architecture requires synchronous communication necessitating the use of a common clock. Having such a facility is difficult to realize in distributed real-time systems. However, it is possible to synchronize the freely running clocks of a DRTS at regular intervals using suitable algorithms. The time-triggered architecture (TTA) described in [8] can be used as a template to implement distributed real-time systems. Each node in this network is allocated a fixed time slot to broadcast any

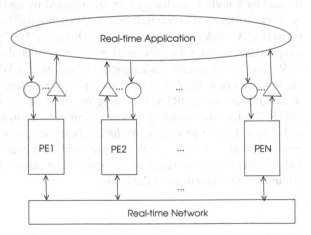

Fig. 3.1 Distributed real-time system structure

message it has, hence providing a guarantee on message delivery. TTA architecture is implemented in many real-time applications such as automotive applications.

In contrast, system must react to external events that arrive sporadically in an *event-triggered system* as we briefly reviewed before. Online, priority-driven scheduling of tasks is needed in such a system. In general, an event-triggered system is more flexible and adaptive to changing system characteristics. However, these systems suffer from significant overheads at runtime due to more complex scheduling algorithms than the time-triggered case. Communication in a distributed real-time system also can be classified as event triggered where communication is started when a *send* command is received, or time triggered in which messages are sent periodically. In general, event-triggered approach is convenient in DRTSs in which unpredictable events from external world occur frequently. On the other hand, time-triggered method is suitable for deterministic DRTSs which have known task sets with characteristics of tasks known beforehand. In the common case, a DRTS will have time-triggered and event-triggered components.

3.2.2 Finite-State Machines

A *finite-state automaton* consists of discrete states and transitions between these states. An FSM is represented by a 6-tuple as follows:

- I is the set of finite inputs,
- O is the set of finite outputs,
- S is a set of finite states,
- $S_0 \subset S$ is the initial state,
- δ is the next state function: $I \times S \rightarrow S$, and
- λ is the output function.

There are two main types of FSMs; the Moore-type FSM and the Mealy FSM. The output depends only on the current state in the former ($\lambda: S \rightarrow O$) and the output is a function of both the input and the current state ($\lambda: I \times S \rightarrow O$) in the latter. An FSM is in the form of a directed graph where states are the vertices of the graph and the directed edges represent transitions between the states. Edges are labeled as $f(input)/output$ to mean this transaction (arc) occurs when the specified input function holds true or simply when the input is received with the specified output generated.

Let us model a simple vending machine that provides fruits worth 20 cents each, which only accepts 5 and 10 cent coins. It has two outputs, R to release the fruit and C to give the change. A person can deposit four 5 cents in a row, and hence we need four 5 cent transactions at least, and there may be 10 cent transactions of course. We will represent 5 or 10 cent inputs as a 2-bit binary number to indicate 5 and 10 cent inputs. For example, 01 means 5 cent and 10 is 10 cent inserted. The output is similarly a 2-bit binary number with the first bit representing the *release* and the second one representing the *change* buttons. The FSM using these specifications with four states is shown in Fig. 3.2.

Fig. 3.2 FSM diagram for
the vending machine

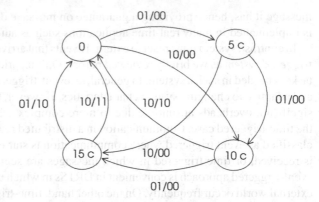

Table 3.1 FSM table for vending machine

	Inputs	01	10
States	0	*act_00*	*act_01*
	1	*act_10*	*act_11*
	2	*act_20*	*act_21*
	3	*act_30*	*act_31*

In order to write a program in C language to implement this FSM, we can use
a simple and effective approach by first forming an *FSM table* that has states as
rows and inputs as columns. Note that we cannot have an 11 input as simultaneous
inserting of 5 and 10 cent coins is not possible. We will represent the 0, 5, 10, and
15 cent states by 0, 1, 2, and 3, respectively. The FSM table for the vending machine
is shown in Table 3.1. Note that we do not need a 20 cent state as reaching 20 cent
deposit means we need to release the fruit and give change or not, and go back to
state 0.

For each table entry, we can define an *action* to be taken and proceeding this way,
the program code is simple to go to the table entry defined by the current state's row
and the current input as the column in the FSM table as shown in the below C code:

```
/*************************************************************
            Vending machine FSM Implementation
*************************************************************/
typedef void (*func_ptr_t)();
func_ptr_t fsm_tab[4][2];
int current_state=0;

 act_00{current_state=1; printf("Total is 5 cents");}
 act_01{current_state=2; printf("Total is 10 cents");}
 act_10{current_state=2; printf("Total is 10 cents");}
 act_11{current_state=3; printf("Total is 15 cents");}
 act_20{current_state=3; printf("Total is 15 cents");}
 act_21{current_state=0; printf("Total 20, release,no change");}
```

```
act_30{current_state=0; printf("Total 20, release, no change");}
act_31{current_state=0; printf("Total 20, release, change");}

void main(){
    int input;
    // initialize
    fsm_tab[0][0]=act_00; fsm_tab[0][1]=act_01;
    fsm_tab[1][0]=act_10; fsm_tab[1][1]=act_11;
    fsm_tab[2][0]=act_20; fsm_tab[2][1]=act_21;
    fsm_tab[3][0]=act_30; fsm_tab[3][1]=act_31;

    current_state=0;
    while(true)
    { printf("Input 0 for 5 cent and 1 for 10 cent");
      scanf("\%d", &input);
      (*fsm_tab[current_state][input])();
    }
}
```

FSMs are simple to use and algorithms for verification and synthesis of distributed real-time systems can be effectively designed. The number of states of an FSM can be very large to easily manage. A DRTS can be modeled by a network of FSMs. Commonly used types of more sophisticated FSMs in DRTSs are as follows.

• Hierarchical FSMs: A state of an FSM can be another FSM which simplifies the design. Hence, activating a state in an HFSM results in activating the FSM associated with that state.
• Concurrent Hierarchical FSMs: This type of FSM is formed by the combination of hierarchy and concurrency arbitrarily. For example, we may have an FSM that can be represented as a state embedded in another FSM, yet it can consist of concurrently running FSMs itself. Commonly, real-time systems can be specified by concurrent hierarchical finite-state machines (CHFSMs) which have both properties of concurrent and hierarchical FSMs.

3.3 Distributed Real-Time Operating Systems and The Middleware

A real-time operating system needs to manage the resources that exist at a node and also has to provide a convenient user interface where applicable as we have reviewed in Chap. 1. Scheduling of tasks, task management, inter-task communication and synchronization, management of input/output, handling of interrupts, and memory management are the main tasks performed by such an operating system. A distributed real-time operating system (DRTOS) has to perform all of the above functions locally at a node and also should provide synchronization and communication among tasks residing at different nodes of the system.

Timeliness is the major requirement from a DRTOS which typically consists of a small real-time kernel, commonly called *microkernel*, replicated at each node of the distributed system that still performs the above-listed functions. Moreover, a distributed real-time kernel now has to coordinate relay of messages over the real-time network. The messages need to be delivered reliably and in time which necessitates support for fast delivery by the network hardware and protocol.

A subset of higher level tasks required from the real-time operating system may be realized by an individual node, that is, the functions needed may be partitioned across the nodes. The higher layers of the operating system work together with the microkernel performing low-level functions close to hardware and the unreplicated components involved in higher level functions.

3.3.1 The Middleware

The middleware layer resides between the operating system and the application in the general sense. It provides functions that are typically extensions of the functions provided by the operating system as well as management of the network. There is a need for this layer rather than embedding the provided functions in the operating system since different applications may need different middleware activities but they all need the basic operating functions. Yet, these procedures are general and may be required by many applications to have a general framework on top of the operating system. Three main middleware functions needed in a distributed real-time system are the clock synchronization, end-to-end scheduling of real-time tasks, and network management.

A clock of a processor commonly uses a crystal oscillator to generate clock pulses. Clocks of nodes in a distributed real-time system will drift apart due to inexactness in clock frequencies as time goes by, to the point where a common time frame will not be possible. Nodes with different clock values will result in erroneous operations in the system. Various clock synchronization algorithms are commonly employed to correct the clock values to a common value at nodes. There are central algorithms in which a master node dictates its clock value to all nodes periodically, or distributed algorithms where nodes exchange messages to reach a common clock value as we will see in Chap. 6.

Another middleware function needed in distributed real-time systems is to provide end-to-end scheduling of tasks. This operation is specific to DRTSs and is needed when an overall deadline is to be achieved while scheduling subtasks of a task.

3.3.2 Distributed Scheduling

An important problem to be handled by the distributed real-time operating system is to schedule tasks among the nodes of the distributed system so that each task meets its deadline and load is fairly distributed. This so-called *distributed scheduling* and *load balancing* is not a trivial task as we will see in Chap. 9. *Static distributed scheduling*

Table 3.2 An example task set

τ_i	C_i	D_i
1	4	12
2	5	20
3	3	15
4	10	60
5	6	30
5	15	60

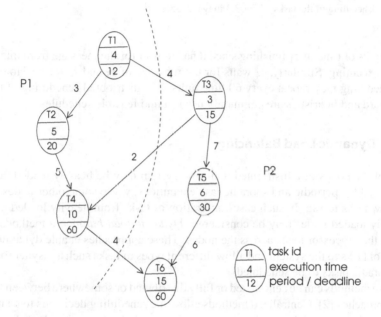

Fig. 3.3 Task graph example

refers to the scheduling of the tasks prior to runtime. One way of achieving this is to have a *task graph* and partition it using some heuristic. Let us consider the hard real-time task set given in Table 3.2 with tasks $\tau_1, \ldots, \tau_6$ having computation times C_i and deadlines D_i.

Let us further assume the communication among these tasks as shown in task graph of Fig. 3.3 where an arrow pointing from a task τ_i to τ_j shows τ_i must finish before τ_j can begin running. Scheduling of these six periodic tasks to two processors P_1 and P_2 can be achieved by partitioning the task graph as shown with a dashed line. While scheduling these tasks, communication between two tasks running in the same processor is considered to have zero cost although inter-processor communication is significant and needs to be taken into account.

A possible scheduling of this set to the two processors using a *Gantt chart* which shows task executions against time is depicted in Fig. 3.4. Note that τ_3 cannot run

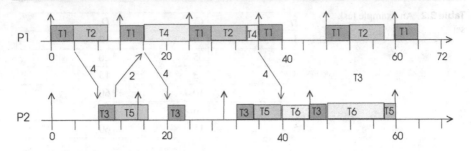

Fig. 3.4 Scheduling of the tasks of Fig. 3.3 in two processors

after 4 units of time of τ_1 finishing since it has to wait for the message from this task to start executing. Similarly, τ_4 waits for the message of τ_3 and τ_6 for τ_4. Note that this scheduling may repeat every 60 time units. The distributed scheduling of tasks is NP-hard and heuristics are commonly used to find feasible schedules.

3.3.3 Dynamic Load Balancing

Some of the nodes of a distributed real-time system may be heavily loaded due to unpredictable aperiodic and sporadic task executions, while some other nodes may have few tasks to run. In such cases, migration of tasks from heavily loaded nodes to lightly loaded nodes may be considered. *Dynamic load balancing* methods aim to even the processor loads across the nodes. These approaches enable dynamically adding of tasks to the system and allow different types of tasks such as asynchronous and sporadic tasks to be scheduled.

These methods can be centralized or fully distributed or somewhere between these two approaches [2]. Centralized methods allow the scheduling decisions to be made at a central node which may become a bottleneck with a large number of nodes. Also, the state information to be kept at the central node may be large. Moreover, the central node is a single point of failure, deficiency of which may cause the system to halt. However, for systems with small number of nodes this approach may be practical and simply to apply.

Decisions are made at each node in distributed load balancing using *sender-initiated* or *receiver-initiated* approaches. A node that finds its load as heavy searches for a light node in the sender-initiated method while a light node searches for a heavy node to receive tasks in the receiver-initiated method.

3.4   Real-Time Communication

General network communications provide best effort service, on the other hand, a real-time communication network should provide a certain level of quality of service (QoS) to the real-time application. The main physical characteristics of a network are as follows:

- *Throughput*: The number of packets that the network can deliver per unit time.
- *Bit Length*: Number of bits traversing the medium simultaneously.
- *Bandwidth*: This is the number of bits that can traverse the channel in a unit of time. The bandwidth of the communication channel should be high enough to provide the required throughput to the application.
- *Propagation Delay*: Time needed for one bit to traverse the communication channel from one end to the other end.
- *Latency*: Time taken to deliver a packet.
- *Jitter*: This is the deviation of a periodic signal from its period.

The main QoS requirements from a real-time network are a bound on the network delay, a bound on the loss rate experienced by the network, and a small blocking probability. In general network communications, *reliability* is the main concern although achieving fast data transfer is clearly preferable. On the other hand, real-time communication requires arrival of messages reliably and in timely manner. Messages may have deadlines within which they must be delivered to the destination application over the *real-time network*. Commonly, provisions for prioritized message passing is provided in a real-time network such that high-priority messages are delivered before the lower ones in the network.

3.4.1   Real-Time Traffic

There exist three main types of traffic in real-time networks (RTNs) [7]:

- *Constant Bit Rate* (CBR) Traffic: The network is utilized at a constant rate, for example, when sensors generate data periodically. Commonly, hard real-time tasks produce CBR traffic using fixed length messages.
- *Variable Bit Rate* (VBR) Traffic: Data transmission over the network varies at different times.
- *Sporadic Traffic*: Network data transfer occurs at bursts typically followed by long periods of no transfers. This type of traffic is a special case of the VBR traffic with the main difference that there is a certain time interval between two sporadic data transfers. A typical example is an alarm occurring in a real-time network.

3.4.2 Open Systems Interconnection Model

The open systems interconnection (OSI) model assumes seven layers of data communication functions over a telecommunication or a computer network. A layer in this model receives commands from and sends responses to the layer above it. This separation of the complicated process of communication simplifies the design and implementation of basic network functions. The layers in this model are as follows:

- *Physical Layer*: The main function of this layer is the transfer of data between the physical medium and the devices.
- *Data Link Layer*: This layer provides point-to-point data transfer between two connected devices. Doing so, it provides mechanisms for flow control and error detection and correction. It consists of two sub-layers: medium access control (MAC) and logical link control (LLC) layers. The MAC layer indulges in controlling network access and the LLC layer is responsible for error detection and flow control.
- *Network Layer*: The main task of this layer is to enable routing of data packets from a source node to a destination node.
- *Transport Layer*: This layer is responsible for correct end-to-end data transfer between two applications. The transmission control protocol (TCP) for connection-oriented and user datagram protocol (UDP) for connectionless communication are the two main protocols in this layer.
- *Session Layer*: Dialog management between two communicating entities is the main task performed in this layer.
- *Presentation Layer*: Syntax and context conversions between two communicating applications are handled in this layer.

The TCP-IP model of communication in the Internet typically uses four layers: the application layer, the transport layer (transport and session layers of the OSI model), the internetworking layer (subset of the OSI network layer), and the link layer (data link and partly physical layers of the OSI model). The seven-Layer OSI model demands considerable overheads. Moreover, many real-time networks are designed for a specific application and use a local network. Hence, the presentation and network layers of the OSI model can be discarded in these networks. Also, short messages are commonly needed to transfer messages in a real-time network making the fragmentation and assembly facilities unnecessary. A collapsed OSI model with three layers is commonly employed in a small RTN as shown in Fig. 3.5 where the application may access the data link layer directly. It is however more common to have another layer typically at transport level to provide a network-independent interface to the application.

3.4.3 Topology

The real-time application commonly consists of processing sites that are within close proximity to each other; hence, the real-time network is typically a local network

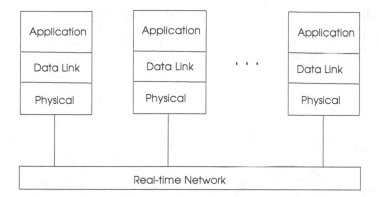

Fig. 3.5 Collapsed OSI model used in a real-time network

consisting of a bus, ring, tree, or a mesh structure as shown in Fig. 3.6. Bus network is very common in general and also in real-time systems. All nodes are connected to a common bus which reduces the wiring; however, a mechanism to share the bus is needed. In the most commonly used method of Ethernet, stations transmit at any time they need and listen to the bus at the same time. If there is a collision, each station involved in it wait for a random amount of time and attempt to transmit again. Bus can be time shared by allocating it to each station for a specific amount of time for communication. In the communication architecture called *token bus*, a token can be circulated among the stations in a predefined order and any station that has the token has the right to transmit.

The ring structure allows passing of a message from one station to another until it reaches its destination. A token is commonly used in this structure to determine the right to transmit. Tree structure can also be used for communication. The worst time to transfer a message from one leaf to another through the root is then twice the depth of the tree. Note that parallel communication through disjoint paths is possible in a tree structure. The mesh structure where each station is connected to all other stations has the best performance as each node is only one hop away from any other at the expense of extensive wiring needed. The required connections would be $n(n-1)/2$ for an n-node network.

The *star network* provides two-hop communication links between any two stations in the network using a total of n links for an n-node network. However, as with all communication systems that have a central node, the central node is a single point of failure and all communications will stop if it breaks down. Moreover, it also presents a bottleneck as all messages are delivered through it. *Wireless communication networks* use radio waves for message transfers employing multi-hop communications, and hence the area must be covered so that there is a connection between any two stations. Wireless sensor networks commonly use a tree structure that is centered around a more powerful root station for communication.

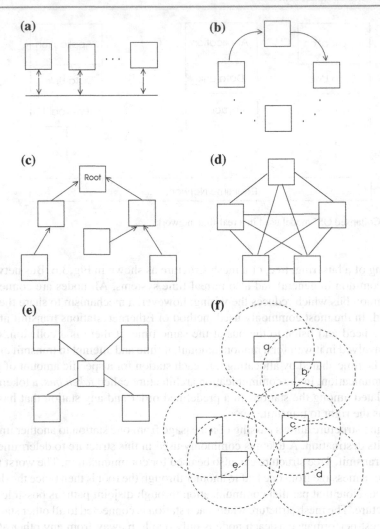

Fig. 3.6 Topology of real-time networks. **a** Bus, **b** Ring, **c** Tree, **d** Mesh, **e** Star, and **f** Wireless architectures. Transmission ranges of nodes in the wireless network are shown by dashed lines. Note that node *a* in the wireless network can communicate with node *f* through nodes *b* and *c* in sequence or just through node *c* using multi-hop communication

3.4.4 Real-Time Data Link Layer

Data link layer manages the flow and error control between two connected devices. Note that the frames processed in the data link of a device can belong to different applications residing in that device. A *transport packet* is the unit of information transferred between two applications in the transport layer of hosts. Data link consists of the MAC and LL layers described below.

3.4.4.1 Medium Access Protocols

The MAC sub-layer mainly deals with the arbitration of the shared communication medium. Three main methods of real-time arbitration at MAC layer are as follows:

- Carrier sense with multiple access/collision detection (CSMA/CD): This is a group of protocols based on sensing the bus before attempting to transmit. A collision detection mechanism is used to detect if two nodes attempt to transmit messages at the same time. When this happens, stations wait for a random amount of time and try to transmit again. This protocol is used in Ethernet and Wi-Fi communications. Due to its nondeterministic behavior, it is not suitable to be used in an RTN. A variation of CSMA/CD that can be used in an RTN is CSMA with collision arbitration (CSMA/CA) where message priorities are imposed. Whenever a collision occurs, the highest priority message is transmitted which provides the determinism needed.
- Time division multi-access (TDMA): As we have outlined before, each station in the networks is allocated a predetermined duration of network usage called *frame* which are typically fixed time slots. This type of protocol is suitable for distributed real-time systems because of its deterministic characteristic. Each station is aware of the time it has right to access to the network. TDMA is commonly used in shared bus distributed real-time systems.
- Token based Communication: Possession of a token provides the right to use the network for message transfer. Token Bus (IEEE 802.4) uses a common bus and Token Ring (IEEE 802.5) uses a ring architecture to transfer the token. A station that has no data to send transfers the token to the next node in the latter in which a token continuously rotates in the network. A token also circulates the network in the Token Bus architecture but the sequence of stations that it can traverse is determined by the software, that is, each station has a predecessor and a successor assigned to it. Fiber distributed data transfer (FDDI, ANSI X3T9.5) has dual counter rotating tokens.

3.4.5 Controller Area Network Protocol

A controller area network (CAN) is a small local network commonly used to connect components of embedded controllers [1]. The length of the CAN network is usually not more than several tens of meters and the speed of communication is commonly 1 Mbps. It was originally started by BOSCH as a network for automotive industry to connect car components such as braking, fuel injection, air conditioning, etc. It is now widely used in automation systems, ships, avionics equipment, and medical equipment.

Modern automobiles may have several tens of electronic control units (ECUs) to control brakes, navigation, engine, etc. CAN uses two wires for the multi-access communication bus resulting in major reduction in the size of wires. CAN is based on CSMA/CA MAC method in which a station that wants to transmit monitors the bus and starts sending its message when the bus is idle. The special bus arbitration

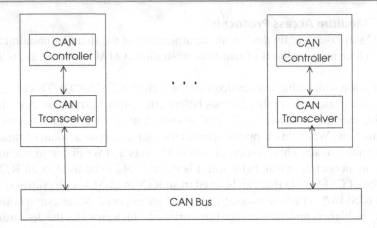

Fig. 3.7 CAN architecture

of the CAN protocol provides efficient handling of the collisions. CAN uses four message types as shown below:

- *The Data Frame*: This is the most common message type to convey data.
- *The Remote Frame*: This frame is used to solicit transmission data from a remote node.
- *The Error Frame*: A node detecting an error in a message sends this frame causing other nodes to send the error frame. The sender of the message then retransmits the message.
- *The Overload Frame*: This frame is used by a busy node to delay transfer of messages.

CAN covers only the physical and data link layers of the OSI model. The physical layer functions such as signaling and medium-dependent interface are handled by the CAN transceiver and the MAC and LLC functions of the data link layer are handled by the CAN controller as shown in Fig. 3.7.

3.4.6   Time-Triggered Protocol

The time-triggered protocol (TTP) is a family of fault-tolerant protocols with small overhead designed by Kopetz and Grunsteidl [5]. TTP has two versions; TTP/A for automotive class A for low-cost soft real-time applications, and TTP/C for automotive class C for hard real-time applications. TTP/A uses master/worker-based bus arbitration in which a master node supervises bus access. TTP employs TDMA to access the network and is based on time-triggered distributed real-time architecture in which each event is defined in a common time base. All clocks of the nodes

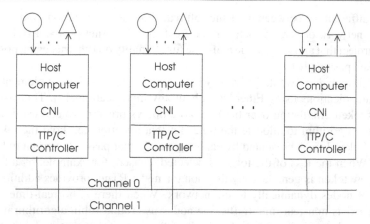

Fig. 3.8 TTP/C network

are synchronized with high precision and messages which are mostly periodic are transferred at predetermined time instants.

TTP/C

TTP/C used for hard real-time systems has a communication network interface (CNI) between the communication controller and the host computer in a network node. The TTP/C communication controller (CC) is the actual interface between the network and the TTP/C station. Each node is assigned a fixed time slot to access the network. The TTP/C network connects nodes by two replicated channels named Channel 0 and Channel 1 as shown in Fig. 3.8. TTP/C can work with the common bus topology or the star or star/bus combination. Clock synchronization between the stations is provided in each TDMA round.

Flexray is a follower of the TTP/C protocol developed by the European car manufacturers. It combines the event-driven and time-driven communication approaches where event-driven communication is possible in some of the TDMA slots.

3.4.7 Real-Time Ethernet

Ethernet as defined in IEEE 802.3 is not a suitable protocol for real-time applications due to its nondeterminism in network access. It is possible that a transmission request in Ethernet is not granted after a limited number of random waits following collisions. Avoiding the collisions is the first step toward generating a deterministic Ethernet.

There are few attempts to transform Ethernet into a real-time protocol. Real-time Ethernet (RETHER), as one of these efforts, was developed at the State University of New York at Stony Brook [11]. In the multi-segment version of this protocol, a hybrid operation is provided in which CSMA/CD mode is activated when there is non-real-time traffic. The stations enter RETHER mode when there is real-time traffic in the network. This mode is realized by circulating a token in the shared bus and a station-reserving bandwidth before may transmit in the cycle. This protocol does not differentiate between hard and soft real-time communications and requires installing network drivers at each station. Real-time Ethernet protocol (RT-EP) is another project that aims to provide real-time functionality on Ethernet using priority-based token passing [6].

RTnet designed at Twente University is yet another approach to implement real-time communications using Ethernet without any modifications [3]. This protocol also uses token circulation over the bus and employs preemptive earliest deadline first (EDF) algorithm to allocate the time slot that a station can hold the token in advance. Token loss is prevented by enabling a node that passes the token to be the monitor. When the loss of the token is detected, caused, for example, by a failing node, a new token is generated by the monitor node. RTnet provides addition and removal of nodes dynamically to the network. When there is no real-time traffic and there exists non-real-time traffic, the nodes use round-robin algorithm instead of EDF for token passing. Main drawback of this protocol is that it also does not differentiate between hard and soft real-time (CBR, VBR, or sporadic) traffic.

3.4.8　Real-Time IEEE 802.11

IEEE 802.11 is a set of MAC and physical layer specifications for wireless communications over a local network [12]. A 802.11 network consists of basic service sets (BSS) that are connected with a distribution system (DS). The 802.11 MAC protocol provides two modes of communication for wireless nodes; the distributed coordination function (DCF) which uses carrier sense multiple-access with collision avoidance (CSMA/CA) and the second function is the point coordination function (PCF) which is used to divide the time between a contention-free period (CFP) and a contention period (CP). Using PCF, a node can transmit data during contention-free polling periods.

IEEE 802.11e is an enhancement to the IEEE 802.11 by proposing prioritization of data, voice, and video transmissions. The enhanced distributed channel access (EDCA) communication mechanism defined in IEEE 802.11e standard is used to support real-time traffic over IEEE 802.11. High-priority traffic has a higher probability of accessing the network using EDCA than a lower priority traffic. IEEE 802.11e provides services for time-sensitive soft real-time applications such as streaming video and voice over Internet protocol (VoIP). Operation at radio frequencies in two ranges; 2.400–2.4835 GHz and 5.725–5.850 GHz are possible with IEEE 802.11e standard, it is however not intended for hard real-time traffic.

There exist few efforts to impose hard real-time traffic over IEEE 802.11. Real-time wireless multi-hop protocol (RT-WMP) is one such protocol based on RT-EP using token circulation with prioritized messages in 802.11 [10]. It supports hard real-time traffic by providing a bounded end-to-end message delay with predetermined duration. Messages are prioritized and multi-hop communication is used to increase network coverage.

3.5 Challenges in Distributed Real-Time Embedded Systems

There are several challenges in the DRTS design. First of all, the general tasks performed by an operating system have to be enhanced in two directions in a DRTS; the operating system is now distributed and also real time. A distributed operating system is envisioned by a user as a single operating system. Realizing this function brings a significant complexity to the operating system design on its own. Yet, we need to provide the basic distributed operating system functions such as inter-task communications and synchronization over the network in real time in a predetermined manner which is a significant task. The middleware in a DRTS is situated between the DRTOS and the distributed real-time application to provide a common set of services to the application. It can therefore be used by many diverse real-time applications. A typical middleware service in a DRTS is the synchronization of clocks.

Fault tolerance is one of the main reasons in implementing a distributed system, whether real time or non-real time. In the real-time case, the cost of a fault may be much higher both in terms of human life and other costs. Therefore, detecting faults such as network failures or software malfunctions and recovering from these faults are fundamental functions to be performed in a DRTS which are commonly realized by the DRTOS and/or middleware. Common sources of faults in a DRTS are the hardware failure of nodes, network failures, transmission delays that exceed the bound, and distributed coordination problems.

As in the single node real-time system, we need to have the tasks meet their deadlines as the most basic and fundamental task. In the distributed case, realization of this function is more difficult and hence a challenge in a DRTS. We will see different approaches that address this problem in Chap. 9. Scheduling tasks with known execution times and deadlines can be performed by partitioning the task graph to the computing nodes available. This function is NP-hard even in the case of non-real-time tasks. Implementation of graph partitioning brings in the additional requirement that the tasks should meet their deadlines. Yet, another problem encountered in a DRTS is the testing which becomes a challenge on its own to handle issues such as environment simulation and fault tolerance, compared to a non-real-time distributed system.

In summary, the main challenges in the design and implementation of a DRTS are the design and implementation of DRTOS, the real-time middleware, fault-tolerant methods, and distributed scheduling.

3.6 Examples of Distributed Real-Time Systems

We will briefly describe the structure of two commonly used distributed real-time systems; a modern car and a mobile wireless sensor network as examples of distributed real-time systems.

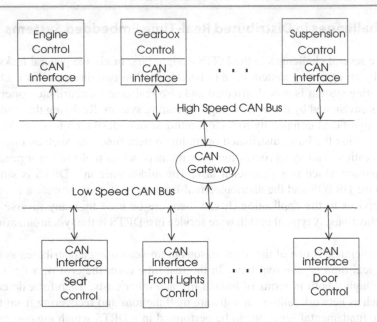

Fig. 3.9 Communication network in a modern car

3.6.1 A Modern Car

A modern car using the CAN protocol is a typical example of a distributed real-time system in which main modules attached to the real-time network have timing constraints. An example of car control system is depicted in Fig. 3.9 in which there are two CAN buses connected by a gateway. The high-speed CAN bus connects modules such as engine, suspension, and gearbox control, whereas the low-speed network is used for communication between front light, seat, and door control units.

3.6.2 A Mobile Wireless Sensor Network

A wireless sensor network (WSN) consists of small autonomous nodes with sensors, limited computational capabilities, and wireless communication using radio waves. A mobile WSN (MWSN) is a sensor network with mobile nodes. MWSNs have numerous applications such as environment monitoring, rescue operations, military surveillance, and healthcare applications [4]. These networks may have various topologies such as tree, mesh, or cluster but commonly hybrid topologies are preferred. The main challenges in an MWSN application are the limited battery lifetime when the nodes are small, and the use of shared medium effectively while the topology changes with time. Routing in these networks uses multi-hop message transfers and has to deal with correct delivery of the messages over a dynamic topology. A MWSN is depicted in Fig. 3.10 with nodes $a, \ldots, i$. A dashed circle around a node

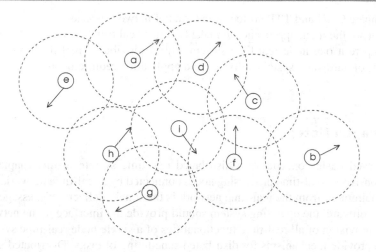

Fig. 3.10 MWSN

shows its transmission range and we can see each node is within transmission range of at least one other node. Hence, network is connected and each node can communicate to any other node in the network. Nodes move in the direction shown by arrows pointed outside and there may be nodes disconnected from the network at the next time instant, for example, node *b* is such a candidate. Providing a connected network at all times is a fundamental problem in an MWSN as shown by this example.

The MWSN nodes can be attached to people, animals, and vehicles for monitoring purposes. Monitoring can be performed periodically using time-triggered method or it can be event based where data is generated when an event occurs. One can have a mixture of these modes as in the general distributed real-time case.

3.7 Review Questions

1. What is the difference between a time-triggered and an event-triggered distributed real-time system?
2. What is an FSM and concurrent hierarchical FSM?
3. What are the differences between a real-time operating system and a distributed real-time operating system?
4. What is the main function of middleware in a distributed system?
5. Compare scheduling and load balancing in a distributed real-time system.
6. What are the common transmission media in general networks?
7. What is a common network architecture for a distributed real-time system?
8. What are the main requirements from a real-time network protocol?
9. What are the main types of traffic in a real-time network?
10. Name the three main methods in real-time MAC protocols.

11. Compare CAN and TTP protocols in terms of network access.
12. What are the main approaches to make Ethernet real time?
13. Compare a one-node real-time system with a distributed real-time system in terms of hardware, system software, and application requirements.

3.8 Chapter Notes

We reviewed basic concepts of a distributed real-time system in this chapter. A DRTS consists of real-time processing nodes connected by a real-time network. The basic requirement from the real-time network is the timely delivery of messages. In terms of software, the operating system should provide the interface to the network with the provision of all real-time functionalities of a single node real-time system, and also provide mechanisms for distributed scheduling of tasks. Distributed static scheduling refers to assignment of tasks to the nodes of the distributed system prior to run time so that each task meets its deadline. For aperiodic and sporadic tasks that are activated at runtime, various dynamic scheduling policies are commonly employed. These methods can be broadly classified as central with a central node managing load transfers and distributed in which each node decides freely to balance its load. Viewed by the initiation of the load transfer, there are sender-initiated and receiver-initiated load balancing approaches. A fundamental requirement from a distributed real-time system middleware is the provision of clock synchronization as all of the tasks need to synchronize at certain points in time over a global time base.

Main challenges in the design and implementation of distributed real-time systems outlined before are selecting a suitable real-time network for the application; choice of a distributed real-time operating system and middleware and writing patches to these software modules as needed by the application. Furthermore, the whole application may be viewed as a single task graph and an offline assignment algorithm to map tasks to the nodes is needed. For unpredictable task activations such as the arrival of an aperiodic task, a dynamic task migration transfer mechanism may be employed.

3.9 Exercises

1. A *parity bit* of a binary number is added to the end to make the total number of bits including the parity bit even or odd. For example, given the 8-bit binary string 10011010, we need to append a 0 for even parity and a 1 for odd parity. A parity checker inputs bits of a binary string one-by-one, and it shows the current parity of stored bits, 0 for odd parity and 1 for even parity. Design a parity checker with *even* and *odd* states using an FSM and implement this FSM in C language.

Table 3.3 An example task set

τ_i	C_i	D_i	Preds	Communication
1	2	12	–	$(1, 2) = 3; (1, 3) = 5$
2	4	15	1	$(2, 4) = 4$
3	5	20	1	$(3, 4) = 2; (3, 5) = 6$
4	9	30	2, 3	
5	12	60	3	

2. A task set $T = \{\tau_1, \ldots, \tau_5\}$ has the following precedence relationship. Task τ_1 has no predecessors, it precedes τ_2 and τ_3; task τ_2 precedes τ_4, τ_3 precedes τ_4 and τ_5 as depicted in Table 3.3. Task τ_4 is the terminal task and has no successors as depicted in Table 3.3. Draw the task graph for this set of tasks first. Given the following characteristics of the tasks in this set, show whether a feasible schedule of the set in two processors is possible or not. Show the schedule using a Gantt chart.

3. Propose a method for prioritized message delivery in a real-time network.

4. A Token Bus protocol is to be realized in a DRTS. Write the basic token circulation MAC layer routine in pseudocode that can be used to receive a token, check whether it can be used, otherwise forward it to the next station.

References

1. CAN bus. http://www.can-cia.org/can/protocol/
2. Erciyes K, Ozkasap O, Aktas N (1989) A semi-distributed load balancing model for parallel rael-time systems. Informatica 19(1):97–109 (Special Issue: Parallel and Distributed Real-Time Systems)
3. Hanssen F, Jansen P, Scholten H. Hattink, T (2004) RTnet: a real-time protocol for broadcast-capable networks, University of Twente, Enschede, 2004. http://www.ub.utwente.nl/webdocs/ctit/1/000000e5.pdf
4. Hayes T, Ali FH (2016) Mobile wireless sensor networks: applications and routing protocols. Handbook of research on next generation mobile communications systems. IGI Global. ISBN 9781466687325, pp 256–292
5. Kopetz H, Grunsteidl G (1993) TTP—a time-triggered protocol for fault-tolerant real-time systems. In: IEEE CS 23rd international symposium on fault-tolerant computing, FTCS-23, Aug 1993, pp 524–533
6. Martinez JM, Harbour MG (2005) RT-EP: a fixed-priority real time communication protocol over standard ethernet. In: Proceedings of Ada-Europe, pp 180–195
7. National Programme on Technology Enhanced Learning. Real-time systems course. Govt. of India

8. Kopetz H (1997) Real-time systems-design principles for distributed embedded applications. Kluwer Academic Publishers
9. Romer K, Matern F (2004) The design space of wireless sensor networks. IEEE Wirel Commun 11(6):54–61
10. Tardioli D, Villaroel JL (2007) Real time communications over 802.11: RT-WMP. In: 2007 IEEE international conference on mobile adhoc and sensor systems, pp 1–11
11. Venkamatrani C (1996) The design, implementation and evaluation of RETHER : a real-time Ethernet protocol. PhD thesis, State University of New York, Stony Brook
12. www.ieee802.org/11

Part II
System Software

Real-Time Operating Systems

4

4.1 Introduction

An operating system resides between the hardware of a computer system and the application by accepting calls from the application and implementing these on the hardware. The hardware in a typical computing system consists of processors, memory, and input/output units. The two main functions of an operating system are to manage various hardware and software resources to be utilized effectively and to hide the details of various hardware and software components from the application and the user. The lowest level of the operating system, which manages bare hardware and various low-level functions is called the *kernel*.

Modern operating systems are based on the concept of a *process* (or a *task*), which is a program in execution. One of the main functions of an operating system is the *task management*, which includes routines for task creation, deleting and scheduling. Facilities for task communication and synchronization are also to be supplied by the operating system. Other resources to be managed by an operating system are the memory, input/output (I/O), and software such as databases. A real-time operating system also has to perform these functions but with always a consideration of the time. It has to be predictable and all services required from the operating system should be delivered with bounded times. For example, there should be a provision of *timed* sending and receiving of messages and synchronization. Also, the application should be able to access certain hardware control functions. A desired property from the real-time operating system is the allowance of choice of different task scheduling policies.

We start this chapter with the review of general operating system concepts and comparison of general operating system and a real-time operating system, highlighting the key points of differences. We then continue with the task, memory, and I/O management functions of a real-time operating system.

© Springer Nature Switzerland AG 2019

K. Erciyes, *Distributed Real-Time Systems*, Computer Communications and Networks, https://doi.org/10.1007/978-3-030-22570-4_4

4.2 A General Operating System Versus a Real-Time Operating System

An operating system consists of software modules that provide a convenient way of accessing hardware and software resources. This software also serves as a manager to efficiently use computer system resources. Although accessing the bare hardware by the user is possible in various computer systems, this provision is not commonly used due to the complexity of writing programs that directly access the hardware. Also, the code that accesses hardware will not be transferrable to another computer in general. The basic resources that are managed by an operating system are the processors, the memory, the input/output devices, and software such as databases.

A real-time operating system has most of the functionalities of a general operating system but it has to provide this operation in a predictable and timely manner. The main characteristics of a real-time operating system are as follows:

- *Interrupt Handling*: Interrupts are generated by external devices and the routine that serves them are called interrupt service routines (ISRs). An ISR runs in high priority mode in a general operating system by preempting the running task. In a real-time system, interrupting a task with a hard deadline may cause missing of the deadline. Hence, the ISR should be short and fast since the prohibition of further interrupts for a long time is undesirable in a real-time system. A common approach is to have the ISR as another task of high priority to be scheduled by the real-time operating system.
- *Synchronization and Communication*: Functions for task creation, destruction, communication, and synchronization should be delivered in a timely manner.
- *Scheduling*: Scheduling is the process of assigning a ready task to the processor. The main concern in a real-time scheduling is that the deadlines of tasks should be met. The context switching time should be kept as minimal as possible which requires efficient ready queue handling procedures. Facilities for periodic, aperiodic, and sporadic tasks scheduling and management should be provided. Various real-time scheduling policies, each with advantages and disadvantages, will be reviewed in Part III in detail.
- *Memory Management*: The memory space of the real-time application should be managed efficiently by the real-time operating system. In order to be predictable, dynamic memory allocation is not preferred in a real-time system to prevent unwanted waiting and uncertainties. All of the needed memory is typically allocated statically at compile time so that waiting for memory at runtime is considerably reduced. However, there is still need for memory management to be able to reuse the space when it is freed. For example, a network driver will pick data from network, will fill an already allocated buffer and deliver it to a higher level protocol. The message will be delivered to the application in the end, which will use the data in the buffer and should return the buffer to the free memory space for further use. If such recycling of free space is not provided, the system will run out of memory space in a short time.

- *Scalability*: A real-time application may vary from a simple one to a very complex one. Hence, the real-time operating system should be scalable to be suitable for a range of applications. In some cases, two or more versions of a real-time operating system are presented to suit small and large applications.

4.3 Task Management

Modern operating systems are designed using the concept of a task, which is an instance of a program executing on a processor. Having a number of tasks provides concurrency and hence improves performance. For example, instead of a single program waiting for a slow I/O device or a resource that is not available, the task that waits for the device or the resource can be stopped and the processor can be assigned to another task. When the resource or the I/O device becomes available, the waiting task can be resumed. A task has a code, and some data part of which may be shared with other processes in the system. Typically, a real-time task has the following attributes:

- *Task Identifier*: This is commonly a unique integer or sometimes a name that identifies a task in the system.
- *State*: The state of a task shows whether it is executing, ready to be executed or waiting for an event.
- *Priority*: Tasks will have different priorities in a real-time system reflecting the order of their execution preferences.
- *Program Counter*: This is a register showing the address of the next instruction to be executed by a task.
- *Registers*: The current values of registers in the processor. These values need to be stored and restored when tasks switch.
- *Stack Pointer*: This is special register that is mainly used for procedure calls. It points to a stack which is a Last-In-First-Out queue to hold return addresses from procedure calls.
- *Period*: For periodic tasks, this is the duration of a period.
- *Absolute Deadline*: The deadline of a real-time task in an absolute time scale.
- *Relative Deadline*: The deadline of a real-time task relative to its arrival time.
- *Memory Pointers*: These are file pointers to the current open files.
- *Input/Output Information*: Current information about I/O devices, whether they are allocated, the times and durations they are used, etc. should be stored.
- *Statistics*: Any statistical information about a task such as time spent waiting for resources and missed deadlines.

All of this information is kept in a data structure called *task control block* (TCB). A running process may block waiting for an event upon which its current data is stored in its TCB by the operating system so that it can be resumed from the last instruction as if it was not blocked, when the event occurs. A task goes through

Fig. 4.1 The basic states of
a process

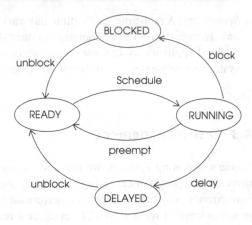

a number of states during its execution which are described in very basic form as
follows:

- *Ready*: A *ready* task has all of the resources it needs to execute except the processor
  which currently is assigned to another task.
- *Running*: The task that is executing on the processor is in the *running* state.
- *Blocked*: A task waiting for a resource or an event is in *blocked* state.
- *Delayed*: A task is delayed for a specific interval of time after which its state is
  changed to *ready* state.

The transitions between these states are depicted in Fig. 4.1. The delayed state is
important in a real-time system to implement periodic tasks which are invoked at
regular intervals.

The *scheduler* is a central component of an operating system that selects a task
to be executed by the processor among a queue (or queues) of ready tasks. There are
various scheduling methods to determine which ready task to assign to the processor.
The lowest part of the scheduler called *dispatcher* is invoked to store the environment of the current task in its TCB and restore the environment of the next task to
be executed from its TCB. Common scheduling policies are first-come-first-served
(FCFS), shortest-job-first (SJF), and priority-based scheduling. Real-time systems
require specific scheduling algorithms which we will review in detail in Part III.

4.3.1 Task Management in UNIX

The UNIX operating system provides various system calls for process management.
The *fork* system call creates a process that has the same code and address space as
the calling process. The created process becomes the *child* of the caller and the caller
is the *parent* of the called. The parent process may learn the identifier of the child
by the returned integer value from the *fork* call.

UNIX provides various process communication schemes; *pipes* are a simple way to pass data between processes. A pipe is a special FIFO file with limited storage. Two basic operations on a pipe is *read* and *write*; a process that wants to read from an empty pipe is blocked. A pipe needs to be created for *read* and *write* operations to enable read/write accesses to it. Pipes may be combined with the *fork* call to pass data between a parent and a child process as shown in the C code below. We have a parent process forking a child, and two arrays of two integers *p*1 and *p*2 are used to store pipe identifiers for bidirectional communication between these processes. The parent writes the contents of the second half of array A to the pipe that has the identifer stored in *p*1 which are read one by one by the child and the total sum is calculated. This sum is returned to the parent via pipe identifier stored in *p*2 which adds its calculated sum of the first half of A to it, and then displays it. Note that the first pipe identifier is used for reading and the second one for writing, and p_1 is used for parent-to-child communication where p_2 is used for the other direction. It is safe to close the unused end of a pipe as in this example. If *pipe* call returns a general error which is -1, we need to exit the program. Array A is allocated in parent and the child but the initialization is not realized in the child since its initial values are passed to the child by the parent.

```c
#include <stdio.h>
#define n   8
int i, c, child_sum, p1[2], p2[2], A[n], total_sum, my_sum=0;

main()

{ if (pipe(p1) == -1) {
    exit(2);
  }
  if (pipe(p2) == -1) {
    exit(2);
  }
  c=fork();   // p2 is from child to parent
  if(c!=0) { // this is parent */
    close(p1[0]);  // close the read end of p1
    close(p2[1]);  // close the write end of p2
    for(i=0;i<n;i++)  // initialize array
        A[i]=i+1;
    write(p[1],&A[n/2],n/2*sizeof(int)); //send half array
    for(i=0;i<n/2;i++)
        my_sum=my_sum+A[i];
    read(p2[0], &child_sum, sizeof(int)); // read sum of child
    total_sum=my_sum + child_sum;
    printf("Total sum is = \%d", total_sum);
  }
  else {   /* this is child */
    close(p2[0]);  // close the read end of p2
    close(p1[1]);  // close the write end of p1
    read(p1[0], &A[n/2], n/2*sizeof(int))); // read half array
    for(i=n/2;i<n;i++)
```

```
      my_sum=my_sum+A[i];
    write(p2[1],&my_sum,sizeof(int));   // send partial sum
    }
}
```

A multitasking operating system allows preemption of a task and assigning one of the ready tasks to the processor. Dividing an application into a number of tasks prevents waste of processor time but the operating system has to provide means for these tasks to synchronize and communicate.

4.3.2 Inter-task Synchronization

The synchronization between tasks is needed in two common cases: to protect shared data from concurrent accesses and when a task is waiting for some action from another task or an interrupt to be able to continue. The *critical section* of a task is the code segment in which it accesses shared data. Execution of a task in a critical section must be *mutually exclusive* of other tasks that share the same data. Let us consider an example where two tasks $T1$ and $T2$ access a shared variable t. They both increment the value of t in a high level language statement such as $t \leftarrow t + 1$ which is typically transferred to three assembly instructions as follows:

```
T1:                     T2:
1. LOAD R1, @t          3. LOAD R2, @t
2. INC   R1             4. INC   R2
6. STORE R1, @t         5. STORE R2, @t
```

where R1 and R2 are registers of the processor. The lines before the instructions show a possible order of execution where $T1$ stops execution after the increment by an interrupt, and the processor is assigned to $T2$ which completes the sequence of instructions. Assume the variable t contains the value of 5 before executions of $T1$ and $T2$. Register $R1$ is loaded with 5, then incremented to 6 and since $T1$ is stopped, its environment including register $R1$ is loaded to its TCB. Task $T2$ now runs, loads register $R2$ with 5, increments it, and stores 6 in t. The processor is assigned to $T1$ at this point and its environment with $R1 = 6$ is restored from its TCB. Finally, $T1$ stores 6 on the variable t which already has this value. Hence, we have incremented the t value once instead of intended twice. This type of situation is known as a *race condition* and should be prevented by the operating system by ensuring there is only one task in its critical section. If tasks in the above example were allowed to finish the three instructions without being interrupted, we would have had a correct operation. One possible way to remedy the race condition such as in the above example is to have the interrupts disabled at the beginning of a critical section and have them enabled when a task finishes execution. However, leaving this control to the application has shortcomings as forgetting to enable the interrupts will stall the whole system. Hence, it is convenient that task synchronization should be handled by the operating system.

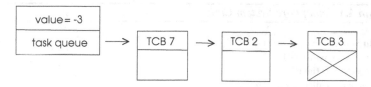

Fig. 4.2 A sample semaphore with three tasks in its queue

Inter-task (or simply task) synchronization can be achieved at hardware level, operating system level, or the application level using a suitable algorithm. We will elaborate on the operating system primitives that provide synchronization. A *semaphore* is a data structure used for task synchronization that consists of an integer and a process queue as depicted in Fig. 4.2.

Two operating *system calls* (or *primitives*) on semaphores are the *wait* and *signal* operations. A simple way to implement a *wait* is to decrement the value of the semaphore, test whether this is less than zero and enqueue the calling task in the semaphore queue if the value is negative as shown in Algorithm 4.1. The *signal* call does the reverse by incrementing the semaphore value and if this value is equal to or less than zero, there must be at least one waiting task at the semaphore queue which needs to be removed from there, its state should be changed to *ready* and placed in a ready queue and finally the scheduler should be invoked so that this task is scheduled immediately or in future, depending on its priority as shown in Algorithm 4.1. Note that *make ready* operation itself may involve calling the Scheduler as we will implement in the sample kernel. The *wait* and *signal* operations should be atomic, that is, they should be executed without being interrupted. A simple way to achieve this indivisibility is to disable interrupts at the beginning of the system call and enable the interrupts at the end as in many other kernel system calls. Note that this approach of disabling and enabling interrupts can be realized at kernel level since this code is written once and tested, but may cause problems at user level as stated.

A real-time system application may need a timed wait system call which checks the value of the semaphore and if this value is 0 or less, it delays itself. Upon waking up, it checks the value of the semaphore again and returns with an error if the value is 0 or negative as shown in Algorithm 4.2.

4.3.3 Inter-task Communication

Tasks need to send data to each other during the progress of their executions. It is the job of the operating system to provide an orderly delivery of messages among tasks. Commonly, data is deposited in buffers called *mailboxes* or *ports*. A mailbox can be implemented by a sender semaphore, a receiver semaphore, and a buffer queue as shown in Fig. 4.3. A mailbox has a limited space to hold pointers for messages and the sender of a message is blocked on the sender semaphore when there is no space in the mailbox. The receiver semaphore queues the tasks that attempt to read a message from the mailbox when the mailbox is empty.

Algorithm 4.1 *Semaphore System Calls*

```
1:
2: procedure WAIT(semaphore s)
3:     s.value ← s.value − 1
4:     if s.value < 0 then
5:         enque the caller in s.queue
6:     end if
7: end procedure
8:
9: procedure SIGNAL(semaphore s)
10:    s.value ← s.value + 1
11:    if s.value <= 0 then
12:        dequeue the first task from s.queue
13:        make the task ready
14:        call the Scheduler
15:    end if
16: end procedure
17:
```

Algorithm 4.2 *Semaphore Wait Call with Time-Out*

```
1:
2: procedure WAIT_TIMED(semaphore s)
3:     if s.value <= 0 then
4:         delay myself n_wait_sem time
5:         if s.value <= 0 then
6:             return NOT_AVAILABLE
7:         end if
8:     end if
9:     s.value ← s.value − 1
10:    return DONE
11: end procedure
12:
```

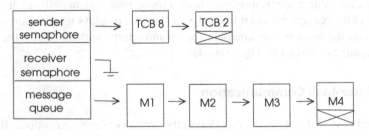

Fig. 4.3 A sample mailbox structure. Two tasks, T8 and T2 waiting for space and four messages M1, M2, M3, and M4 are queued to be received in the mailbox

We can see that there are four messages in the example mailbox meaning there will be four receive operations by tasks from this mailbox without getting blocked on the receiver semaphore. Two tasks, $T8$ and $T2$ are waiting for free space on the mailbox and will deposit their messages when two tasks retrieve messages $M1$ and $M2$ from the mailbox. The two main system calls on a mailbox are *send* and *receive* operations as shown in Algorithm 4.3.

Algorithm 4.3 *Mailbox System Calls*

1:
2: **procedure** SEND(*mailbox mb, message msg*)
3:     *wait*(*mb.send_sem*)
4:     **enqueue** message in *mb.message_queue*
5:     *signal*(*mb.receive_sem*)
6: **end procedure**
7:
8: **procedure** RECEIVE(*mailbox mb, message msg*)
9:     *wait*(*mb.receive_sem*)
10:     **dequeue** message *msg* from *mb.message_queue*
11:     *signal*(*mb.send_sem*)
12: **end procedure**
13:

In a real-time environment, a task should not be blocked for an undetermined time on a mailbox semaphore. Therefore, *send* and *receive* routines with time-out can be provided. A *receive* with time-out will then wait for a message, and a *send* with time-out will wait for free message space for the specified time intervals, commonly in milliseconds. If the required space or message is not present after time-out, these routines will return with error as shown in Algorithm 4.4. Otherwise, the *send* and *receive* routines work as the blocking versions.

4.3.4 UNIX Interprocess Communications

UNIX provides various interprocess communication methods including the pipes we have already reviewed. *Message queues* and *shared memory* are the two other basic communication schemes provided by UNIX. Shared memory is protected by semaphores with message queues being a more general communication approach among processes that do not need to have a common memory. We will briefly review message queues as this method is general and also, we will be using these routines to simulate network communications while implementing the experimental kernel. A message queue is created by the system call *msgget* with the following format.

```
int msgget( key_t key, int msgflg );
```

which returns a nonnegative integer that identifies the queue and the low order bits of the *msgflg* are used for access permissions to the queue. The key variable is used

---

Algorithm 4.4 *Mailbox System Calls with Time-Out*

```
1:
2: procedure SEND_TOUT(mailbox mb, message msg)
3:    if mb.send_sem.value <= 0 then
4:        delay myself n_wait time
5:        if mb.send_sem.value <= 0 then
6:            return NO_SPACE
7:        end if
8:    end if
9:    wait(mb.send_sem)
10:    enqueue message in mb.message_queue
11:    signal(mb.receive_sem)
12:    return DONE
13: end procedure
14:
15: procedure RECEIVE_TOUT(mailbox mb, message msg)
16:    if mb.receive_sem.value <= 0 then
17:        delay myself n_wait time
18:        if mb.receive_sem.value <= 0 then
19:            return NO_MSG
20:        end if
21:    end if
22:    wait(mb.receive_sem)
23:    dequeue message from mb.message_queue
24:    signal(mb.send_sem)
25:    return DONE
26: end procedure
27:
```

---

as a seed to create a specific message queue. Sending a message to a message queue is performed by the *msgsend* routine described below

```
int msgsnd(int msqid, const void *msgp, size_t msgsz, int msgflg);
```

where *msqid* is the integer returned from the *msgget* system call, *msgp* is the address of the message to be sent, and *msgsz* is the size of the message. The *msgflg* field specifies the action when the queue is full; IPC_NOWAIT means the call returns without waiting and with error set, and setting this flag to 0 means the sender is blocked until the message is sent or the process catches a signal or the queue is removed from the system. Receiving a message has a similar format but additionally, the message type is specified to allow selective receiving,

```
int msgrcv (int msqid, void *msgp, size_t msgsz, long
                        msgtyp, int msgflg);
```

which returns the number of bytes received when successful. The *msgflg* field set to IPC_NOWAIT means the receiving process should not be blocked if there is no message and MSG_NOERROR specifies that the messages should be truncated if the space allocated is not big enough. UNIX message queues require copying of data from user memory space to kernel space while sending and receiving which may result in significant overheads for large volumes of data. Also, broadcasting is difficult since reading from a message queue removes data from the queue.

4.4   Threads

A process or a task may consist of a number of lightweight subprocesses called *threads*. A thread is the smallest unit of execution by the processor. Threads are used for their low context switching times to improve performance since only a small amount of the environment, typically consisting of the register set, private data memory and private memory, needs to be stored and restored. The rest of memory is global to all threads of the same task. Threads also provide effective resource sharing and improved responsiveness, for example, a thread responsible to serve a request will respond to that request faster than the task in terms of activation overheads. A multi-threaded application can also be scheduled conveniently in a multicore/multiprocessor system.

4.4.1   Thread Management

Two types of threads used are the *kernel threads* and the *user threads*. A kernel thread is scheduled by the kernel and is known to the kernel, whereas a user thread is only valid in the user space. Scheduling a kernel thread requires context switching as in the task scheduling, however, it has about an order of magnitude less overhead than task context switch due to the size of data stored in its *thread control block*. A user thread is not known to the kernel hence has to be managed in the user space by a thread library which provides all thread management utilities such as creation and deletion of threads and inter-thread synchronization and communication. User thread context switching has much less overhead in context switch than a kernel thread. Blocking of a kernel thread by the kernel will not stop other kernel threads of the task but a user thread blocking on a resource will block the whole task. It is therefore convenient to employ kernel threads when these are involved in frequent I/O operations which may block them. On the contrary, the user threads are commonly used in applications that require intense computations with rare I/O instructions commonly for ease of design and modularity. Threads are typically used for parallel processing applications such as Web interface where many users may attempt to reach a server or in a multiuser operating system.

4.4.2   POSIX Threads

Portable Operating System Interface (POSIX) standard (IEEE 1003.1c) [4] is an attempt to standardize UNIX for diverse applications. Many contemporary operating systems comply to POSIX and hence provide portability for the application code.

This interface standard defines the application programming interface (API) and other utilities to provide a common interface for different versions of UNIX and other operating systems. The threads in POSIX are called *Pthreads*. We will elaborate specifically on POSIX thread management utilities. The subroutines that comprise Pthreads API can be grouped into following [5]:

- *Thread Management*: Routines that are defined directly on threads such as creating, deleting.
- *Mutexes*: A mutex is used for mutual exclusion. The related routines are used for mutual exclusion while accessing a shared resource.
- *Condition Variables*: Routines used for synchronization between threads that share a mutex.
- *Synchronization*: Routines used for synchronization via read/write locks, barriers and semaphores.

The following function in POSIX interface is used to create a thread and invoke it:

```
pthread_create(&t, NULL, my_thread, (void *));
```

This function creates a thread with default attributes, stores its identifier in the variable t for further access and invokes the thread function at address my_thread without passing any parameters to it. Commonly, an increasing counter starting from 1 is used to send counter value to each thread created so that a thread can use this value as its own identifier instead of the identifier assigned by the operating system.

4.4.2.1 Mutual Exclusion

All the threads of a task share the address space of the task which should be protected against race conditions as in the shared memory synchronization of tasks. This problem can be overcome by having the critical section of a thread to execute mutually exclusive of other threads that may use the same shared variable. The mutual exclusion variables in POSIX are declared as $mutex\_t$ variables and locked with the system call $mutex\_lock$ and unlocked by calling $mutex\_unlock$ as shown in the following example where two threads $T1$ and $T2$ share the variable $shared$ and perform safe operations on it by locking and unlocking the mutex variable m. This lock will be set by the first thread that accesses the variable and the other thread will wait, until this thread finishes execution of the critical section and resets the lock to enable the waiting thread. Note that the output of this program could be 6 or 4 depending on which thread executes first.

```
#include <stdio.h>
#include <pthread.h>

int shared=1;
pthread_mutex_t m;
```

```
void *T1(){
          pthread_mutex_lock(&m);
          shared=shared+1;
          pthread_mutex_unlock(&m);
          pthread_exit(NULL);
}

void *T2(){
          pthread_mutex_lock(&m);
          shared=3*shared;
          pthread_mutex_unlock(&m);
          pthread_exit(NULL);
}

main(){
 pthread_mutex_init(&m, NULL); // initialize mutex
 pthread_t t1,t2;
 pthread_create(&t1, NULL, T1, NULL);
 pthread_create(&t2, NULL, T2, NULL);
 pthread_join(t1, NULL);
 pthread_join(t2, NULL);
 printf(" shared = \%d", shared);
}
```

Note that the main program creates the threads and wait for them to finish by calling *pthread_join* function which is synchronized by the *pthread_exit* call by the threads. The main program in the code above is the *main thread* which controls and monitors the creation and termination of threads.

4.4.2.2   Synchronization

There are cases when a thread needs to wait for a signal from another thread. This situation is different than waiting for a lock for mutual exclusion, and semaphores are commonly used for this purpose. The POSIX interface provides semaphore types, and *wait* (*sem_wait*) and *signal* (*sem_post*) operations on the semaphores as shown in the following example with two threads $T1$ and $T2$ where $T1$ writes the contents of the first integer array *data*1 to a shared integer variable *shared* one by one and activates $T2$ after each writing. $T2$ copies the variable contents to the second array *data*2 and signals $T1$ to continue. Two semaphores $s1$ and $s2$ are needed and without using them, there may be overwritten data to the variable *shared*. Thread creation and exit code are not shown. The semaphore $s1$ is initialized to 1 and $s2$ to 0 for correct operation.

```
#include <pthread.h>
#include <semaphore.h>

int shared;
int data1[10]={...},data2[10];
sem_t s1,s2;
```

```
T1(){
  int i;
  ...
  for(i=0; i< 10; i++)
    { sem_wait(&s1);
      shared=data1[i];
      sem_post(&s2);
    }
  ...
}

T2(){
  int i;
  ...
  for(i=0; i< 10; i++)
    { sem_wait(&s2);
      data2[i]=shared;
      sem_post(&s2);
    }
  ...
}
```

POSIX interface specifies condition variables of the type *pthread_cond_t* and two routines *pthread_cond_wait* and *pthread_cond_signal* to wait on a condition variable and signal it. Inter-thread synchronization using condition variables is possible, however, semaphores provide a general structure that can be used for synchronization and communication.

4.4.2.3 Communication

Inter-thread communication using shared variables protected by mutual exclusion variables, and synchronization using semaphores or condition variables is possible. However, a higher level mechanism for transfer of messages between threads is commonly pursued for ease of programming and a simple interface to the data transfer routines. This functionality is not provided by the POSIX interface and hence, we will describe a simple and novel message passing interface in C language that uses statically allocated message queues protected by semaphores as in [2]. A *buffer* is the basic data unit to be passed between the threads and a *message queue* is a contiguous block of buffer pointers that uses a *read index* and a *write index*. Two semaphores (*fullsem* and *emptysem*) provide synchronization for sending to and receiving from a message queue and a mutual exclusion variable (*msgquemut*) is used to protect the indices against concurrent accesses. The message queue structure is defined in the below C code.

```
/ * * * * * * * * * * * * * * * * * * * * * * * * * * * * * * * * * * * * * * * * * * * * * * * * * * * * * * * * * *
                 message queue data structure
* * * * * * * * * * * * * * * * * * * * * * * * * * * * * * * * * * * * * * * * * * * * * * * * * * * * * * * * * * */


#define MSGQUE_SIZE 10
#define N_MSGQUES 10
#define ALLOCATED 1
#define ERR_MSGQUEEMPTY -1
#define ERR_MSGQUEFULL -2

typedef struct msgque *msgqueptr_t;
typedef struct msgque{
                    int state ;
                    int msgque_size;
                    int read_idx;
                    int write_idx;
                    sem_t fullsem;
                    sem_t emptysem;
                    pthread_mutex_t msgquemut;
                    bufptr bufs[MSGQUE_SIZE]; } msgque_t;
```

This structure is initialized to store a number of messages with the semaphores as below

```
/ * * * * * * * * * * * * * * * * * * * * * * * * * * * * * * * * * * * * * * * * * * * * * * * * * * * * * * * * * *
                 initialize a message queue
* * * * * * * * * * * * * * * * * * * * * * * * * * * * * * * * * * * * * * * * * * * * * * * * * * * * * * * * * * */


int init_msgque(msgqueptr mp) {
                    int msgqueid;
                    mp->state=ALLOCATED;
                    mp->msgque_size=MSGQUE_SIZE;
                    sem_init(&mp->fullsem,0,0);
                    sem_init(&mp->emptysem,mp->msgque_size,0);
                    pthread_mutex_init(&fp->msgquemut,0);
                    mp->read_idx=0;
                    mp->write_idx=0;
                    return(msgqueid); }
```

Sending to the message queue is realized by waiting on the sender semaphore and then storing the message.

```
/ * * * * * * * * * * * * * * * * * * * * * * * * * * * * * * * * * * * * * * * * * * * * * * * * * * * * * * * * * *
                 receive a buffer from a message queue
* * * * * * * * * * * * * * * * * * * * * * * * * * * * * * * * * * * * * * * * * * * * * * * * * * * * * * * * * * */
bufptr recv_msgque(msgqueptr mp){
                    bufptr bp;
                    sem_wait(&fp->fullsem);
                    pthread_mutex_lock(&fp->msgquemut);
```

```
bp=fp->bufs[fp->read_idx++];
fp->read_idx MOD=fp->msgque_size;
pthread_mutex_unlock(&fp->msgquemut);
sem_post(&fp->emptysem);
return(bp); }
```

Message reception by the caller is performed by waiting on the receiving semaphore and then removing the message as below. Note that this structure is similar to mailbox-based communication among tasks.

```
/*************************************************************
              send a buffer to  a message queue
*************************************************************/

bufptr send_msgque(msgqueptr mp, bufptr bp){
                bufptr bp;
                sem_wait(&fp->emptysem);
                pthread_mutex_lock(&fp->msgquemut);
                fp->bufs[fp->write_idx++]=bp;
                fp->write_idx MOD=fp->msgque_size;
                pthread_mutex_unlock(&fp->msgquemut);
                sem_post(&fp->fullsem);
                return(bp); }
```

4.5 Memory Management

Management of memory is a fundamental function required from a real-time or a general operating system. Memory allocation can be broadly classified as *static* and *dynamic allocation*.

4.5.1 Static Memory Allocation

Static memory allocation is the allocation of memory space needed by the application at compile time. Memory allocation and deallocation are not needed at runtime in this method. In its simplest form, the programmer allocates space in the form of structures such as variables and arrays in the code. Static memory management approach is suitable for real-time systems as it is deterministic and hence, this method is frequently employed in real-time operating systems. A major problem with static allocation of memory is that we may have reserved space more or less than needed.

4.5.2 Dynamic Memory Allocation

Dynamic memory allocation involves the allocation of memory space as needed at runtime. This method uses memory space efficiently than the static one, however, it can result in significant overheads during execution and worse, the space may not be available at runtime. Dynamic memory allocation can be performed either manually or automatically. The programmer controls when memory is allocated or deallocated in the manual dynamic memory management using calls typically provided by the programming language. For example, the C programming language provides a function *malloc* to reserve space at runtime. The allocated storage should be deallocated after use by the program by a call (*free* in C language). In automatic memory allocation, the programming language or an extension to it provides an automatic memory manager commonly called the *garbage collector* which collects unused memory and recycles it.

4.5.3 Virtual Memory

The user application does not directly access the physical memory, instead, it envisions a *virtual memory* that is very large reserved for itself. With virtual memory, we do not need the whole address space of a task to be resident in memory. The needed code/data can be transferred from the disk when needed.

The virtual memory is divided into *pages* and the physical memory is divided into *frames* of fixed length. A virtual page is mapped to a physical frame, in reality, the virtual memory is treated as a cache for pages stored on the disk. The virtual addresses are translated into physical memory addresses by the memory management unit (MMU) which is commonly implemented in hardware to be fast. The MMU maintains a *page table* to perform this translation. A referenced page may not be in the memory causing a *page fault*, then, that page is transferred from the disk to memory. A place for that page must be found by removing an existing page. Similar problems we have seen in caches are seen in virtual memory management. We need to decide on which page to remove using a page replacement algorithm such as least recently used (LRU) or first-in-first-out (FIFO). Virtual memory is not generally employed in real-time systems due to the overhead in accessing the pages.

4.5.4 Real-Time Memory Management

Static memory allocation is convenient for a real-time system since it is deterministic as stated before. However, a real-time application may have dynamically changing sizes of data to process, making it difficult to judge the size of the needed space at design or compile time. A commonly employed approach in a real-time operating system then is to allocate memory partitions called *memory pools* (or *buffer pools*) with fixed size of blocks in each partition and to dynamically allocate buffers from these pools at runtime. This intermediate strategy between a static and dynamic

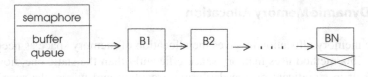

Fig. 4.4 A buffer pool

memory allocation allows the unused memory to be used by tasks that need to process large amounts of data. Figure 4.4 depicts a buffer pool of N buffers.

Two basic system calls on a pool are *get*, which returns the address of a free buffer from the pool and *put*, which returns the address of a used buffer to the pool as shown in Algorithm 4.5. A semaphore at each pool is used to block the calling task when there are no free buffers in the pool. The task that returns a buffer to the pool should signal the semaphore to unblock any task that may be waiting for a free buffer.

Algorithm 4.5 *Buffer Pool System Calls*

1:
2: **procedure** GET(*pool p, buffer_ptr bp*)
3:     *wait*(*p.semaphore*)
4:     **dequeue** the first free buffer address from the pool to *bp*
5:     *return*(*bp*)
6: **end procedure**
7:
8: **procedure** PUT(*pool p, buffer_ptr bp*)
9:     **enqueue** *bp* to the pool *p*
10:     *signal*(*p.semaphore*)
11: **end procedure**
12:

4.6 Input/Output Management

The I/O system consists of I/O devices, device controllers, and the software associated with the devices. Data acquisition systems and A/D are examples of I/O devices in real-time systems and a device controller is an electronic circuit that is used as an interface between a device and the processor. For example, a program will write some value to a register in a device controller to perform a write operation to a device.

4.6.1 Interrupt-Driven I/O

An *interrupt* is an event that stops the processor from executing its current task. An interrupt can be from an external source or internal such as from a timer when it overflows. An interrupt service routine (ISR) is then invoked which processes the request and finally the interrupted task is resumed. *Interrupt latency* is the time interval between the generation of the interrupt and the activation of the ISR. A processor with many pipeline stages needs to reset its stages before it can start the ISR and hence, the interrupt latency time may be significant. In *vectored interrupt* processing, the device that causes the interrupt supplies the processor with the address of the ISR to activate over the system bus. The processor typically goes to the same memory address in any interrupt in *non-vectored* interrupt. External control lines may be used by the polling method to detect the source in the latter as we have seen in Chap. 2. In a real-time system with many I/O devices, vectored interrupt would be preferred since polling many devices at each interrupt would be time consuming. Some processors have two or more interrupt input lines with different priorities at hardware level.

A real-time system should provide priority levels for interrupts such that a lower priority ISR can be preempted by a higher priority ISR. An ISR can call other functions and typically unblocks a blocked task to activate it. An ISR should not wait on a semaphore or a mutex since it may get blocked disturbing servicing of any further interrupts from the same source. The types of interrupt handlers in general are as follows:

- *Non-preemptive Interrupt Handler*: Interrupts are disabled until the current ISR finishes execution, hence, the running ISR cannot be preempted. This method is not suitable for real-time systems which have interrupts of varying priorities.
- *Preemptive Interrupt Handler*: Interrupts are enabled during the running of an ISR enabling nesting of interrupts. However, enabling is usually performed after some vital action such as storing important data is performed. This way, the ISR can be considered to consist of a critical section after which the interrupts are enabled and a noncritical section. There are no priorities among ISRs and size of stack should be considered carefully in this method as this grows with nested interrupt number.
- *Prioritized Interrupt Handler*: The ISRs have varying priorities and preemption of an ISR by a higher priority ISR only is allowed. This approach is the most convenient one for a real-time system as it reflects the external processing in such an environment.

In conclusion, the ISR of a real-time system should be as short as possible, the interrupt latency should be kept small and the ISRs should be prioritized.

4.6.2   Device Drivers

A *device driver* is basically a software module that hides the details of the device from the operating system. A device driver has code specific to the device and it acts on the *device controller* which is a hardware component that controls the device. A disk driver, for example, will issue commands to the disk controller to control the movement of disk head and transfer data. A *device control block* (DCB) is a data structure that holds information about a device. Commonly, the address of a device

Fig. 4.5 Device I/O
hardware and software
structure

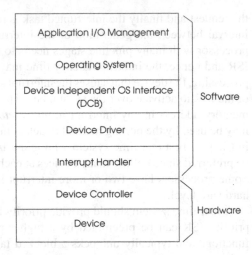

driver is kept in its DCB and as far as the operating system is concerned, call to a
device driver from the DCB is sufficient to perform the required operation.

A device driver initializes the device it is associated with, interprets commands
from the processor such as *read* from or *write* to device, handles interrupts, and manages data transfers. This layered software architecture for I/O handling is depicted
in Fig. 4.5. The hardware components are the device itself and the device controller
which reside on the device. When the application software issues a command on the
device such as read a number of bytes, the operating system calls the related routine
from the DCB of the device. This abstraction allows various devices to be interfaced
in a similar manner to the operating system as in the UNIX operating system.

4.7 A Survey of Real-Time Operating Systems

We will briefly review commonly used real-time operating systems in this section
with emphasize on their structures, methods for inter-task synchronization and communication, and interrupt handling. This selected set of operating systems have a
wide spectrum of applications from avionics to industrial control.

4.7.1 FreeRTOS

FreeRTOS is a simple real-time operating system kernel for embedded systems,
distributed under the open-source MIT license [7]. FreeRTOS is written in C for the
most part with only a small part in assembly language of the target architecture. It
supports many processor architectures including Intel and ARM. As commonly found
in many other real-time kernels, it provides thread and task management routines

including semaphores for synchronization. It occupies small memory space typically between 6 and 12 KB and works fast. It has been used extensively in small real-time applications using microcontrollers and microprocessors.

4.7.2   VxWorks

VxWorks is a real-time multitasking operating system designed for embedded applications by Wind River Systems [6]. It has been used in many applications including various NASA Mars rover projects and Mars Pathfinder. It supports MIPS, Intel, Power, and ARM architectures. VxWorks is based on processes which may consist of a number of tasks. VxWorks provides system level, task service, task control, network management, and I/O functions. Priority-based scheduling and round-robin scheduling are the two main methods to assign tasks to the processor. Priority is determined on resource requirements such as time and memory. Inter-task communication and synchronization is handled by semaphores, queues, and pipes. Basic signal routines are also provided as a form of software interrupt between the processes. VxWorks provides various POSIX compliant APIs. Network management is provided by the socket interface or the remote procedure call to communicate with a peer application.

4.7.3   Real-Time Linux

Linux is a free operating system, which has most of the basic features of the Unix operating system. Real-time Linux (RTLinux) was developed as an extension to the Linux operating system with added real-time capabilities to make it predictable [1]. The main sources of unpredictability in the Linux operating system is the scheduler which is optimized for best throughput rather than being predictable; and interrupt handling and virtual memory management.

RTLinux is constructed as a small real-time kernel that runs under Linux and as such, it has a higher priority than the Linux kernel as depicted in Fig. 4.6. Interrupts are first handed to the RTLinux kernel and the Linux kernel is preempted when a real-time task becomes available. When an interrupt occurs, it is first handled by the ISR of the RTLinux which activates a real-time task resulting in the scheduler to be called. Passing of interrupts between the RTLinux and Linux is handled differently by different versions of RTLinux.

The Linux operating system handles device initialization, any blocking dynamic resource allocation and installs the components of the RTLinux [11]. The scheduler and the real-time FIFOs are two core modules of RTLinux and rate monotonic and earliest deadline first scheduling policies are provided by RTLinux. The application interface includes system calls for interrupt and task management.

Fig. 4.6 RTLinux structure
Adapted from [11]

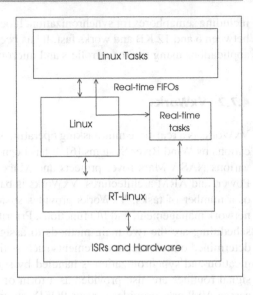

4.8 Review Questions

1. What are the general states of a task?
2. When a task enters an operating system, what should be its initial state and why?
3. What is the main procedure for creating a new task in UNIX?
4. Is it possible to have positive integer values in both of the semaphores of the mailbox structure of Fig. 4.3? Give reason.
5. How does a thread improve responsiveness of a real-time application? Give an example.
6. Compare the advantages and disadvantages of static and dynamic memory management. What type of memory management should be preferred in real-time systems?
7. What is a buffer pool?
8. Compare interrupt-driven I/O with polled I/O.
9. What is a device driver and a device controller? State briefly the interface between them.
10. What makes RTLinux real time?

4.9 Chapter Notes

We provided a dense review of general and real-time operating system concepts in this chapter. An operating system is a resource manager that also provides a convenient interface to the application to perform various tasks on a computer system. Modern operating systems are centered around the concept of a task (process) to be able to perform the above functions efficiently. An operating system manages tasks, memory, and input/output system. There are various books on operating systems that describe these functions in much more detail including [8–10].

A real-time operating system has to provide the functions of a general operating system with significant differences. First, interrupt handling in a real-time operating system requires preemption and possibly multilevel interrupt management. Scheduling has to be performed considering the meeting the deadlines of tasks as the most important criteria. Memory allocation is commonly static eliminating indeterminism encountered in dynamic memory allocation. The real-time operating systems should be scalable to meet the needs of a very diverse set of real-time applications.

4.10   Programming Exercises

1. Write the pseudocode of a clock ISR which is awaken at every clock tick, decrements the value of the clock tick at the front of the *delta queue* and if the front task has zero tick, it awakens it and calls the scheduler.
2. Provide the *block*, *unblock*, and *delay* routines in pseudocode.
3. Provide the data structure for a counting semaphore in C language.
4. Provide the *wait* and *signal* procedures in pseudocode for a binary semaphore.
5. Provide a modification to the *wait* operation on a binary semaphore so that waiting in a queue is possible when the resource is not available.
6. Provide a static array of task control blocks (TCBs) such that each TCB points to the next one in the array initially. The array head structure holds the front and rear of the free TCBs. Write a routine in C language to allocate a free TCB from this array.
7. Show how a buffer pool structure can be realized using a static array and a header.
8. Write a C program in which two threads *T1* and *T2* are formed. Thread *T1* reads a string of characters from the keyboard until end of line is encountered for five times, and sends the read string to *T2* using the message queue structure of Sect. 4.4.2.3.
9. Three POSIX threads *T1*, *T2*, and *T3* work concurrently. *T1* periodically receives data from a heat sensor and T2 receives periodically inputs from a pressure sensor and *T1* and *T2* write data to a commonplace of 2 bytes, 1 byte for data type (heat or pressure) and 1 byte for value. *T3* is then activated, reads the type and value and displays it on the screen. Write a C program with short comments using POSIX threads to have a correct operation. Do not show error handling when system calls are made.

References

1. Barabanov M, Yodaiken V (1997) Real-time Linux. Linux J
2. Erciyes K (2013) Distributed graph algorithms for computer networks. Springer, App. B
3. Erciyes K (1989) Design and realization of a real-time multitasking kernel for a distributed operating system. PhD thesis, Ege University
4. http://pubs.opengroup.org/onlinepubs/9699919799/
5. https://computing.llnl.gov/tutorials/pthreads/
6. https://www.windriver.com/products/product-notes/PN_VE_6_9_Platform_0311.pdf
7. FreeRTOS open source licensing. www.freertos.org/a00114.html
8. Silberschatz S, Galvin PB (2012) Operating system concepts. Wiley. ISBN-10: 1118063333
9. Stallings W (2017) Operating systems, internals and design principles, 9th edn. Pearson. ISBN 10: 0-13-380591-3
10. Tanenbaum A (2016) Modern operating systems. Pearson. ISBN-10: 93325
11. Yodaiken V (1999) The RTLinux manifesto. In: Proceedings of 5th Linux conference

Design of an Experimental Distributed Real-Time Kernel

5

5.1 Introduction

We will describe the design and implementation of a distributed real-time kernel (DRTK) simulator to run on the UNIX operating system that can be used to test real-time system software design strategies. We will be extending functions of DRTK throughout the book and finally we will design and implement a simple distributed real-time application using DRTK; an environment monitoring system using a wireless sensor network. This kernel is structured like the UNIX operating system and similar to [1] in various respects as we will see but is much smaller and simpler than UNIX and also has an internal structure with real-time characteristics similar in structure to the one described in [2]. Having a simulator running on top of a non-real-time operating system will possibly result in losing some of the real-time characteristics of the DRTK. However, our aim is to show the internal architecture and low-level details of a distributed real-time kernel rather than implementing a real-time kernel with favorable performance.

The kernel consists of a lower and a higher module. The lower module has the Scheduler at the bottom and time and interrupt handling, task state managing, and input/output management as higher layers. The higher module consists of task synchronization and communication, high memory management, and task management layers. We will extend this kernel to a distributed one by adding network communication layer in Chap. 6. The DRTK routines should be atomic, possibly implemented by disabling and enabling interrupts, which is omitted in the code for simplicity.

5.2 Design Strategy

We will implement real-time application tasks as POSIX threads and will use UNIX message queue interprocess communication to simulate network communication

© Springer Nature Switzerland AG 2019

K. Erciyes, *Distributed Real-Time Systems*, Computer Communications and Networks, https://doi.org/10.1007/978-3-030-22570-4_5

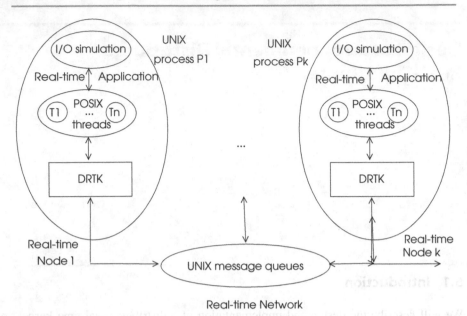

Fig. 5.1 Distributed real-time system development using DRTK

between the nodes of the real-time system as shown in Fig. 5.1, where each real-time node is a UNIX process.

The following attributes of DRTK make it suitable to be used as a real-time kernel simulator.

- We have real-time scheduling policies to meet deadlines of tasks. This property is the most important aspect of a real-time system.
- All of the data structures are statically declared at compile time and hence we do not have unpredictable memory allocation delays.
- Task synchronization and communication system calls have *waiting-with-time-out* versions that may be used in a real-time system.
- The code of DRTK is kept as short and as simple as possible.

5.3   Lower Level Kernel Functions

The low-level DRTK consists of the Scheduler, the Clock Management and Interrupt Handling, the Task State Management, and the Input/Output Management layers as shown in Fig. 5.2. The reason for choosing such a hierarchy is that whenever there is a change of state of the current running task, there may be a need to dispatch a new task, for example, when the current task gets blocked waiting for an event. We

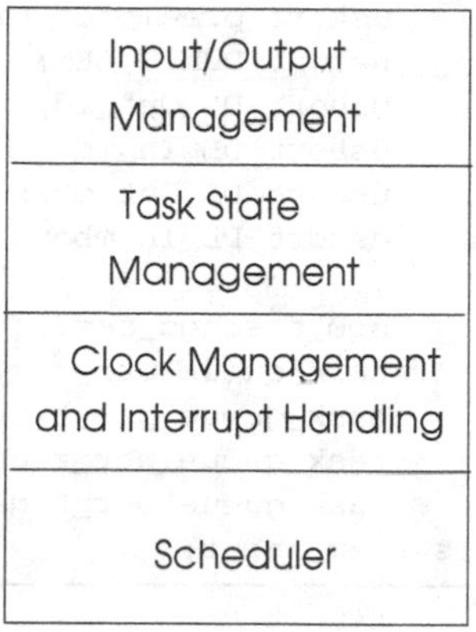

Fig. 5.2 The lower layers of
DRTK

Input/Output Management
Task State Management
Clock Management and Interrupt Handling
Scheduler

will first describe data structures and queue procedures to be used in DRTK and then these layers in a bottom-up approach.

5.3.1   Data Structures and Queue Operations

The key data structure is the system table (*System_Tab*), which contains various system parameters such as the maximum allowed number of tasks, maximum number of pools, etc. It is a global data structure in file "system_data.h", which can be accessed by all tasks residing at the real-time node. It also contains identifiers of system tasks. Some of these tasks will be implemented in Chap. 6 but we include them here for completeness. Local data associated with a module is commonly included in its header file.

```
/*****************************************************************
                      System Table
*****************************************************************/
typedef unsigned short ushort;
typedef int TASK;

typedef struct {
    ushort this_node;          // node identifier
    ushort N_TASK=120;         // data structure limit values
    ushort N_SEM=300;
    ushort N_MBOX=100;
    ushort N_MBOX_MSG=100;
    ushort N_GROUP=50;
    ushort N_GROUP_MEM=30;
    ushort N_POOL=20;
    ushort N_POOL_MSG=400;
    ushort N_DEVICE=120;
    ushort N_DATA=1024;
    ushort N_INTERRUPT=32;
    ushort DELAY_TIME=10;
```

```
    ushort preempted_tid;    // preempted task id
    ushort DUNIT_LEN;
    ushort DL_Out_id;        // system task identifiers
    ushort DL_In_id;         // used for distributed RTS
    ushort DL_Out_mbox;
    ushort DL_In_mbox;
    ...
    sem_t sched_sem;         // POSIX semaphore for Scheduler
  task_queue_ptr_t system_que_pt;
  task_queue_ptr_t realtime_que_pt;
  task_queue_ptr_t user_que_pt;
  task_queue_ptr_t delay_que_pt;
}system_tab_t;
```

5.3.1.1   Data Unit Type

We have the *data unit* as the basic communication structure to be exchanged between the tasks locally or over the network. It contains data as well as transport layer and data link layer headers to be used in network communications, which will be extended when network communication is reviewed. This structure is declared in the header file "data_unit.h" as shown below. We also declare a data unit queue type that will be used as a queue of data units along with pointer types.

```
// data_unit.h
/****************************************************************
                         data unit type
****************************************************************/
typedef struct {
    ushort sender_id;
    ushort receiver_id;
    ushort type;
    ushort seq_num;
}TL_header_t;

typedef struct {
    ushort type;
    ushort length;
    ushort sender_id;
    ushort receiver_id;
}MAC_header_t;

typedef struct *data_ptr_t{
    TL_header_t TL_header;
    MAC_header_t MAC_header;
    int type;
    char data[N_DATA];
    ushort MAC_trailer;
    data_ptr_t next;
}data_unit_t;
```

```
typedef data_unit_t *data_unit_ptr_t;

typedef struct {
    int state;
    data_unit_ptr_t front;
    data_unit_ptr_t rear;
}data_unit_que_t;

typedef data_unit_que_t *data_que_ptr_t;
```

We have the following functions to get and store a data unit to a data unit queue contained in the file "data_unit_que.c" and also a function to check whether data queue is empty or not. The codes for these functions are contained in Appendix B.

- *enqueue_data_unit(data_que_ptr_t   dataque_pt,   data_unit_ptr_t   data_pt)*:
  Enqueues the data unit shown by *data_pt* to the data unit queue at address *dataque_qpt*.
- *data_unit_ptr_t   data_pt   dequeue_data_unit(data_que_ptr_t   dataque_pt)*:
  Dequeues the first data unit from the data unit queue at address *dataque_qpt* and returns its address at *data_pt*.

5.3.1.2 Task Control Block Data Type

The key data structure of a task is the *task control block* (TCB) that carries all vital information about it. It is contained in the "task.h" header file and has the contents shown below. We show the parameters needed only for now but we will extend it as we extend the structure of DRTK.

```
// task.h
/*****************************************************************
                        task states
*****************************************************************/

  #define    RUNNING     1
  #define    READY       2
  #define    BLOCKED     3
  #define    DELAYED     4
  #define    SUSPENDED   5
  #define    RECEIVING   6


/*****************************************************************
                        task types
*****************************************************************/

  #define    SYSTEM      1
  #define    REALTIME    2
  #define    PERIODIC    3
  #define    APERIODIC   4
  #define    SPORADIC    5
```

```
#define    USER        6

/*****************************************************************
                    task related data
*****************************************************************/

#define    VOIDTASK    0
#define    TIMEQUANT   10
#define    N_REGS      32
```

The *task_tab* type is also declared as we will have all of the task control blocks kept in this table, which is statically allocated at compile time. We also need to define task queue structure to be used in scheduling and inter-task synchronization.

```
/*****************************************************************
                    task control block
*****************************************************************/

typedef struct *task_ptr{
    ushort init_address;     // initial task address
    ushort ISR_address;      // interrupt service routine address
    int REGS[N_REGS];        // register values
    ushort tid;               // task identifier
    ushort type;             // task type
    ushort priority;         // task priority
    ushort active_prio;      // active priority
    ushort state;            // task state
    sem_t sched_sem;         // semaphore to wait
    ushort allocation;       // task control block allocation
    ushort mailbox_id;       // mailbox of the task
    ushort group_id;         // group task belongs
    ushort group mailbox;    // group mailbox identifier
    ushort n_children;       // number of children in a tree
    ushort mailbox_access[N_MBOX]; // access to other mailboxes
    ushort abs_deadline;     // absolute deadline of a task
    ushort rel_deadline;     // relative deadline of a task
    ushort delay_time;       // time in ticks a task is delayed
    ushort executed;         // time task has executed
    ushort wcet;             // worst case execution time
    ushort n_predecessors;   // number of predecessors
    ushort n_successors;     // number of successors
    ushort predecessors[N_PREDS]; predecessor task identifiers
    ushort successors[N_SUCCS]; successor task identifiers
    task_ptr next;           // pointer to next task in queue
}task_control_block_t;

typedef task_control_block_t *task_control_block_ptr_t,
        *task_ptr_t;

typedef task_control_block_t task_tab_t[System_Tab.N_TASK];
```

```
typedef struct {
    int state;
    int n_task;
    task_control_block_ptr_t front;
    task_control_block_ptr_t rear;
}task_queue_t;
```

We have the following functions to manipulate the task queues contained in the file "task_que.c". The codes for these functions are given in Appendix B.

- *enqueue_task(task_que_ptr_t taskque_pt, task_ptr_t task_pt)*; Enqueues the task with pointer *task_pt* to the task queue at address *taskque_pt*.
- *task_ptr t task_pt dequeue_task(task_que_ptr_t taskque_pt)*; Dequeues the first task from the task queue at address *taskque_pt* and returns the address of the task at *task_pt*.
- *insert_task(task_que_ptr_t taskque_pt, task_ptr_t task_pt)*; Inserts the task with pointer *task_pt* to a position in the task queue at address *taskque_pt* according to its priority.

5.3.2   A Multi-queue Scheduler

The scheduler of an operating system is one of the most frequently executed code segments, which has to be short and efficient as it affects the whole performance of the system. We will design real-time scheduling algorithms for DRTK in the next part. For now, we will consider a general scheduler that works with three queues; a non-preemptive first-come-first-served (FCFS) system queue for system tasks, a preemptive priority-based real-time queue for hard real-time tasks and a preemptive FCFS user task queue as depicted in Fig. 5.3. System tasks such as a network input and output tasks serve real-time application tasks, therefore they are considered as higher priority tasks than the real-time ones. The job of the scheduler in this version of the DRTK is to select the highest priority task according to the following logic:

1. If the caller (current task) is running and is a system task, next task to run is the caller, no preemption is needed here.
2. If the caller is running and is a real-time task, check system task queue. If it is not empty, remove the first task from this queue and select it as the next task to run. If system task queue is empty, check the real-time task queue. If this is not empty, compare the priority of the first task with the caller. If the calling task has a higher priority, next task to run is the caller, otherwise remove the first task from the real-time queue and select it to run by preempting the current task.
3. If the caller is blocked, then select the first task from one of the queues by checking them in turn from highest priority system task queue to the lowest user task queue.

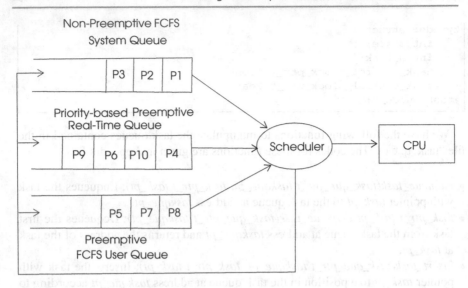

Fig. 5.3 The operation of the sample Scheduler

We need an interface to the UNIX at this point so that it should be possible to stop the running thread. In our implementation, the *Scheduler* is a system task represented by a POSIX thread that is activated by the system call *Schedule*, called by task state management routines when there is a change in the state of a task. This routine simply saves the preempted task identifer of the preempted task and signals the semaphore the *Scheduler* thread is waiting as shown below in the file "schedule.c". Note that if the scheduler was implemented as a function rather than a task or if it did not modify the current task identifer, we did not need to save the identifer of the preempted task identifier in the system table. This routine also stops the caller from execution by waiting on its scheduling semaphore. Note that this is needed even in the case of continuing with the current task.

```
//schedule.c
/*******************************************************************
                    Schedule system call
********************************************************************/

void Schedule(){

    if (current_pt->state==RUNNING)
        System_Tab.preempted_tid=current_tid;
    sem_post(&(System_tab.schedule_sem));
    sem_wait(&(task_tab[current_tid].schedule_sem));
}
```

The Scheduler thread that is always running and is activated only by its semaphore as shown below. Note that we need to store the registers and other environment variables of the current task at each entry to the *Scheduler* and restore the registers and other environment variables of the selected task before exiting this task in a real application. The *current_tid* global value is set to the task identifier of the next task to run by the scheduler. We did not specify how a user task can be preempted, we will assume it gets into state *delayed* when its time quanta expires.

```c
//schedule.c
/****************************************************************
                    Scheduler Task
****************************************************************/

TASK Scheduler(){

    task_ptr_t task_pt;
    task_queue_ptr_t task_qpt;
    ushort next_tid;

while(TRUE) {
  sem_wait(&(System_Tab.sched_sem));
  //store(current_tid); store registers and environment

 switch(task_tab[current_tid].state) {
  case BLOCKED:
  case DELAYED:
        taskq_pt=System_Tab.system_que_pt;
        if (taskq_pt->front!=NULL) {
          task_pt=dequeue_task(System_Tab.system_que_pt);
        }
        else {
          taskq_pt=System_Tab.realtime_que_pt;
          if (taskq_pt->front!=NULL) {
            task_pt=dequeue_task(System_Tab.realtime_que_pt);
            next_tid=task_pt->tid;
          }
        }
        else {
          taskq_pt=System_Tab.user_que_pt;
          task_pt=dequeue_task(System_Tab.user_que_pt);
        }
        next_tid=task_pt->tid;
        break;
  case RUNNING:
        if(task_tab[current_tid].type==SYSTEM)
          next_tid=current_tid;
        else if (task_tab[current_tid].type==REALTIME){
          taskq_pt=System_Tab.realtime_que_pt;
          if (taskq_pt->front != NULL)
            task_pt=taskq_pt->front;
            if(task_tab[current_tid].priority>task_pt->priority)
              next_tid=current_tid;
              else {
                task_pt=dequeue_task(System_Tab.realtime_que_pt);
                next_tid=task_pt->tid;
              }
        }
        else next_tid=current_tid;
        break;
 }
    current_tid=next_tid;
    current_pt=&(task_tab[current_tid])
    //restore(current_tid); restore registers and environment
    sem_post(&(task_tab[current_tid].sched_sem));
}
```

5.3.3   Interrupt Handling and Time Management

Interrupts are key to the operation of a real-time system whether time-triggered or event-triggered. An interrupt service routine (ISR) is invoked when an interrupt occurs, which typically unblocks a blocked task or updates a variable, etc. For fast response, an ISR in a real-time system may consist of low and high levels. The low-level is executed upon an interrupt and the higher level is commonly delayed by leaving it to be scheduled when the processor is not busy. We will have one-level ISR structure in DRTK for simplicity. The ISR table (*ISR_tab*) contains the addresses of the ISRs as in a vectored interrupt method as shown below.

```
//isr.h
/*****************************************************************
                        ISR Table
*****************************************************************/
typedef struct {
    int (*func_ptr)() [System_Tab.N_INTERRUPT];
}ISR_tab_t;
```

The general ISR (*ISR_Gen*) can then be coded as below. It receives the interrupt number and activates the necessary ISR from the *ISR_tab* as indexed by this number.

```
//isr.c
/*****************************************************************
                general interrupt handler
*****************************************************************/

int ISR_Gen(ushort int_num){
    (*ISR_tab[int_num])();
    return(DONE);
}
```

5.3.3.1   The Delta Queue

An important function required from a real-time kernel is the delaying of a task as needed in periodic activation. A *delta queue* is a queue of delayed tasks that are sorted with respect to their delayed times. Any task in this queue is delayed by the sum of delay times of all tasks preceding it in the queue. This way, decrementing only the delay time of the first task in the queue provides decrementing delay times of all tasks as shown in Fig. 5.4.

The *Delta_Queue* data structure is implemented as a task queue. We need to provide an insert operation to this queue by the *insert_delta_queue(task_id, n_ticks)* function details of which are shown in Appendix B. The delay time of the caller may be smaller than the delay time of the first task in this queue and in this case, the caller is placed in front of the queue and the delay of the previous front of the task is set to its previous value minus the caller delay value. The delay value of the caller may be greater than the first task in the queue. In this case, we need to advance in the queue until the first task where the total delay is greater than the delay of the caller.

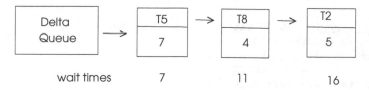

Fig. 5.4 Delta_Queue implementation

5.3.4 Task State Management

The main functions in this module are the management of task transitions between states. We assumed a multilevel scheduler for now but will describe different schedulers appropriate for real-time systems in the next part. All of these system calls are in file "task_state.c".

We may want to change the priority of a task in a case such as a lower priority task holding a resource, which results in a higher priority task waiting and missing its deadline. This problem is known as *priority inversion* as we will review in Chap. ??.

```
// task_state.c
/**********************************************************
                change priority of a task
**********************************************************/

int change_prio(ushort task_id, ushort prio){

    task_ptr_t task_pt;

    if (task_id < 0 || task_id >= System_Tab.N_TASK)
        return(ERR_RANGE);
    task_pt=&(task_tab[task_id]);
    task_pt->priority=prio;
    return(DONE);
}
```

A task attempting to acquire a resource may get blocked which is realized by the *block_task* system call. We have provided facility to allow blocking of a task by another task as shown below. For this reason, the scheduler is entered only if the caller is the currently executing task, that is, it is blocked due to a resource not available. Stopping of the current task is realized in the *Schedule* function as stated. This function signals the semaphore the *Scheduler* is waiting to make it runnable.

```
/**********************************************************
                block a task
**********************************************************/

int block_task(ushort task_id){

    task_ptr_t task_pt;
```

```
    if (task_id < 0 || task_id >= System_Tab.N_TASK)
        return(ERR_RANGE);
    task_pt=&(task_tab[task_id]);
    task_pt->state=BLOCKED;
    if (task_id==current_tid){
        Schedule();
    }
}
```

Unblocking a task is achieved by changing its state to *ready* and enqueuing/inserting its task control block to one of the ready queues based on its type. We check the parameter *sched* to determine whether the *Scheduler* should be awaken or not. This is desirable since the caller of this function may want to continue without having to be rescheduled as in the case of a hard real-time task that must meet its deadline and hence should not preempt.

```
/******************************************************************
                        unblock a task
******************************************************************/
int unblock_task(ushort task_id, ushort sched){

    task_ptr_t task_pt;

    if (task_id < 0 || task_id >= System_Tab.N_TASK)
        return(ERR_RANGE);
    task_pt=&(task_tab[task_id]);
    task_pt->state=READY;
    switch task_pt->type {
        case SYSTEM    : enqueue_task(&systask_que, task_pt);
                         break;
        case REALTIME  : insert_task(&realtime_que, task_pt);
                         break;
        case USER      : enqueue_task(&user_que, task_pt);
    }
    if (sched==YES)
        Schedule();
}
```

A task may need to delay itself for a specific number of clock ticks. The system call for this operation is *delay_task*, which is typically called by a periodic hard real-time task to delay itself for its period.

```
/******************************************************************
                        delay a task
******************************************************************/

int delay_task(ushort task_id, ushort n_ticks){

    if (task_id < 0 || task_id >= System_Tab.N_TASK)
```

```
                return(ERR_RANGE);
           insert_delta_queue(task_id, n_ticks);
           task_tab[task_id].state=DELAYED;
           if (task_id==current_tid)
                Schedule();
    }
```

5.3.4.1 The Time Interrupt Service Routine

The Time Interrupt Service Routine (*Time_ISR*) is invoked at every clock tick, which
we will set to 100 μs. The delay times of tasks are specified as clock ticks, and hence,
it suffices for this ISR to provide decrementing of all clock ticks of waiting tasks at
Delta Queue by decrementing the clock tick value of the fist task in this queue. It
then checks whether the first task has a zero delay value and if this is so, it dequeues
this task from the queue and makes it ready for execution as shown below by calling
the *unblock* routine, which will be described in the next section.

```
// isr.c
/******************************************************************
                          Time ISR
******************************************************************/

int Time_ISR(){

    task_ptr_t task_pt;

    while(TRUE){
        sleep(System_Tab.N_CLOCK);
        task_pt->delay_time--;
        if (task_pt->delay_time==0){
        task_pt=dequeue_task(delta_queue);
        unblock_task(task_pt->tid,YES);
        }
    }
}
```

5.3.5 Input/Output Management

Each input/output device is represented by a *device control block* data structure as in
UNIX. This data structure mainly has pointers to the device drivers associated with
the input/output unit. In our implementation, we have function pointers to read/write
a block of bytes from/to devices in the data structure *dev_cont_block* in the file
"device.h". The device table, *dev_tab*, has *dev_cont_block* data structure as entries.

```
//device.h
/****************************************************************
                    device data structure
****************************************************************/

typedef struct dev_block{
    ushort state;
    func_ptr_t (*read_pt)(char* read_addr, ushort n_byte);
    func_ptr_t (*write_pt)(char* write_addr, ushort n_byte);
    dev_block *dev_pt;
}   dev_cont_block_t, *dev_ptr_t;

typedef dev_cont_block_t *dev_cont_block_ptr_t;
typedef dev_cont_block_t dev_tab_t[System_Tab.N_DEVICE];
```

We form the *dev_cont_block* data structure by the *make_dev* system call as depicted below. It searches the device table for an allocated device control block entry, starting from the first entry.

```
// device.c
/****************************************************************
                      make a device
****************************************************************/

int make_device(char* read_addr, char* write_addr){

    ushort i;
    for(i=0; i<System_Tab.N_DEVICE; i++)
        if (dev_tab[i].state!=ALLOCATED){
            dev_tab[i].state=ALLOCATED;
            dev_tab[i].read_pt=read_addr;
            dev_tab[i].write_pt=write_addr;
            return(i);
        }
    }
    return(ERR_NO_SPACE);
}
```

Reading a block of bytes from a device is performed by the *read_dev* system call, which basically activates the read driver of the device through the device control block as shown below

```
/****************************************************************
                     read from a device
****************************************************************/
int read_dev(ushort dev_id, char *data_pt, ushort n_bytes) {

    dev_cont_block_ptr_t dev_pt;
```

```
       if (dev_id < 0 || dev_id >= System_Tab.N_DEVICE)
           return(ERR_RANGE);
       dev_pt=&(dev_tab[dev_id]);
       (*dev_pt->read_pt)(data_pt,n_bytes);
       return(DONE);
}
```

Writing a block of bytes to a device is done similarly by the *write_dev* system call as below

```
/********************************************************************
                       write to a device
********************************************************************/
int write_dev(ushort dev_id, char *data_pt, ushort n_bytes) {

       dev_cont_block_ptr_t dev_pt;

       if (dev_id < 0 || dev_id >= System_Tab.N_DEVICE)
           return(ERR_RANGE);
       dev_pt=&(dev_tab[dev_id]);
       (*dev_pt->write_pt)(data_pt,n_bytes);
       return(DONE);
}
```

Deleting a device from the system is performed by the *delete_dev* system call, which de-allocates the device control block allocated to the device as shown in the code below

```
/********************************************************************
                       delete a device
********************************************************************/
int delete_dev(ushort dev_id){

       if (dev_id < 0 || dev_id >= System_Tab.N_DEVICE)
           return(ERR_RANGE);
       dev_tab[dev_id].state=NOT_ALLOC;
       return(DONE);
}
```

5.4   Higher Level Kernel Functions

Upper level kernel functions make use of the lower layers, which are data management routines, task state management, interrupt and time handling, and device management. The upper layer calls may be grouped into task synchronization, task communication, and high memory management as depicted in Fig. 5.5.

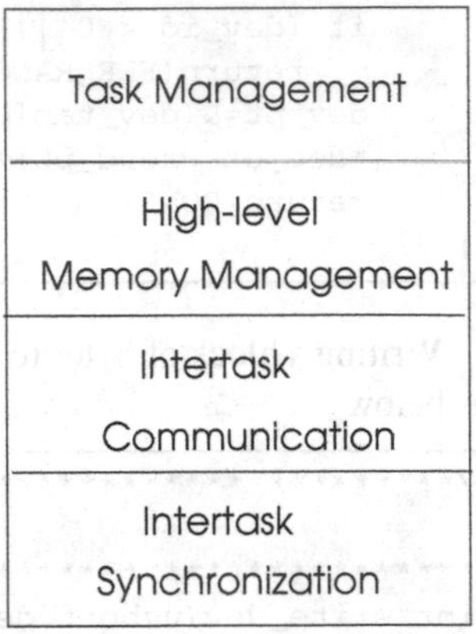

Fig. 5.5 The higher layers of
the sample kernel

Task Management
High-level Memory Management
Intertask Communication
Intertask Synchronization

5.4.1 Task Synchronization

A *semaphore* is the basic synchronization object in DRTK. The semaphore table
semaphore_tab is an array of semaphores created at system initialization. The call
make_sema is used to allocate an element of this table, initializing a semaphore is
by *init_sem* and resetting it to free all of the tasks waiting on it is achieved by the
reset_sem system calls as below

```
// semaphore.h
/***************************************************************
                  semaphore data structure
***************************************************************/

typedef struct{
    int state;
    int value;
    task_queue_t task_queue;
} semaphore_t;

typedef semaphore_t  *semaphore_ptr_t;
typedef semaphore_t semaphore_tab_t[N_SEM];
```

The first function we need is the initialization of a semaphore by the *init_sem*,
which sets its value to the given input parameter and resets the front and rear task
queue pointers of the semaphore.

```
// sempahore.c
/***************************************************************
                     initialize a semaphore
***************************************************************/

int init_sem(ushort sem_id, int val) {
```

```
        if (sem_id < 0 || sem_id >= System_Tab.N_SEM)
            return(ERR_RANGE)
    sem_tab[sem_id].value=val;
    sem_tab[sem_id].task_queue.front=NULL;
    sem_tab[sem_id].task_queue.rear=NULL;
    return(DONE);
}
```

Allocation of a semaphore in the semaphore table is performed by the *make_sem* system call, which also initializes it by the *init_sem* call. Note that *init_sem* function can also be used by the application.

```
/*****************************************************************
                        make a semaphore
*****************************************************************/

int make_sema(int value) {

    ushort i;
    for(i=0; i<System_Tab.N_SEM; i++)
        if (sem_tab[i].state!=ALLOCATED){
            sem_tab[i].state=ALLOCATED;
            init_sem(sem_tab[i],value);
            return(i);
        }
    return(ERR_NO_SPACE);
}
```

Waiting on a semaphore is done by the *wait_sema* system call shown below. The value of the semaphore is decremented and if this is less than or equal to zero, the caller is blocked.

```
/*****************************************************************
                        wait on a semaphore
*****************************************************************/
int wait_sema(ushort sem_id){

    semaphore_ptr_t sem_pt;

    if (sem_id < 0 || sem_id >= System_Tab.N_SEM)
        return(ERR_RANGE)
    sem_pt=&(semaphore_tab[sem_id]);
    sem_pt->value--;
    if (sem_pt->value < 0){
        enqueu_task(sem_pt->task_queue);
        block(current_tid);
    }
    return(DONE);
}
```

A real-time task may want to check whether a resource is available and it may need to continue without the resource to meet its deadline when the resource is not available. We have a system call *notwait_sema* for this purpose where the caller is not blocked in any case. The caller needs to check the returned value to determine the course of action.

```
/****************************************************************
                 check a semaphore without waiting
 ****************************************************************/
int notwait_sema(ushort sem_id){

    semaphore_ptr_t sem_pt;

    if (sem_id < 0 || sem_id >= System_Tab.N_SEM)
        return(ERR_RANGE)
    sem_pt=&(semaphore_tab[sem_id]);
    if (sem_pt->value < = 0)
      return(ERR_RES_NOTAV);
    else
      sem_pt->value--;
    return(DONE);
}
```

Signalling a semaphore is performed by the *signal_sema* system call in which the value of the semaphore is incremented and if this is less than or equal to zero, the first waiting task at the semaphore queue is freed by dequeuing it and then unblocking it to make it eligible to be activated by the scheduler. Different than the usual signalling of a semaphore, we have the parameter *sched* passed to this function to determine whether the scheduler should be invoked with this operation. This feature may be useful in a case when a hard real-time task has a tight deadline. Note that this approach has a shortcoming since providing the application with this facility may have effect on system performance by starving the waiting tasks at the semaphore.

```
/****************************************************************
                   signal a semaphore
 ****************************************************************/

int signal_sema(ushort sem_id, int sched){

    semaphore_ptr_t sem_pt;
    task_ptr_t task_pt;

    if (sem_id < 0 || sem_id >= System_Tab.N_SEM)
        return(ERR_RANGE)
    sem_pt=&(semaphore_tab[sem_id]);
    sem_pt->value++;
    if (sem_pt->value <= 0){
        task_pt=dequeu_task(sem_pt->task_queue);
        unblock_task(task_pt->task_id, sched);
    }
}
```

Our last call in this module is the resetting of a semaphore by the *reset_sema* system call which is needed when a semaphore needs to be deleted from the system. Deleting a semaphore is simply freeing its data structure from the semaphore table and resetting it.

```
/****************************************************************
                      reset a semaphore
 ****************************************************************/

int reset_sema(ushort sem_id, sched){

    semaphore_ptr_t sem_pt;
    task_ptr_t task_pt;

    if (sem_id < 0 || sem_id >= System_Tab.N_SEM)
        return(ERR_RANGE)
    sem_pt=&(semaphore_tab[sem_id]);
    if (sem_pt->value < 0) {
      for(i=sem_pt->value; i<0; i--) {
        task_pt=dequeu_task(sem_pt->task_queue);
        unblock_task(task_pt->task_id,NO);
      }
    }
    sem_pt->state=NOT_ALLOC;
}
```

5.4.2 Task Communication

Task communication is mainly handled indirectly by the use of *mailbox* objects. The mailbox data structure has a data unit queue, a state, a sender semaphore to block the caller when the mailbox is full and a receiver semaphore to block the caller when there are no messages in the mailbox as shown below

```
// mailbox.h
/****************************************************************
                  mailbox data structure
 ****************************************************************/

typedef struct{
    int state;
    ushort next;
    semaphore_t send_sem;
    semaphore_t recv_sem;
    data_unit_que_ptr_t queue;
} mailbox_t;

typedef mailbox_t *mailbox_ptr_t;
typedef mailbox_t mbox_tab_t[System_Tab.N_MBOX];
```

Allocation of a mailbox is performed by the *make_mbox* system call in the usual sense, by obtaining a free entry from the mailbox table (*mailbox_tab*). The sender mailbox is initialized to the maximum number of messages that can be stored in the mailbox, and the receiver semaphores is reset since there are no messages available.

```c
// mailbox.c
/****************************************************************
                     make a mailbox
****************************************************************/

int make_mailbox() {

    ushort i;
    for(i=0; i<system_tab.N_MBOX; i++)
      if (mailbox_tab[i].state!=ALLOCATED){
        mailbox_tab[i].state=ALLOCATED;
        sem_id=make_sema(System_Tab.N_MBOX_MSG);
        mailbox_tab[i].send_sem=sem_id;
        sem_id=make_sema(0);
        mailbox_tab[i].recv_sem=sem_id;
        mailbox_tab[i].queue->front=NULL;
        mailbox_tab[i].queue->rear=NULL;
        return(i);
      }
    return(ERR_NO_SPACE);
}
```

Sending a message to a mailbox is achieved by the *send_mbox_notwait* when the sender does not want to wait for a reply to ensure the message is received by the receiver. However, the caller has to perform a *wait* on the sender semaphore of the mailbox to ensure there is space to deposit the message.

```c
/****************************************************************
                 send a message to a mailbox
****************************************************************/

int send_mailbox_notwait(ushort mbox_id, data_unit_ptr_t data_pt){

    mailbox_ptr_t mbox_pt;

    if (mbox_id < 0 || mbox_id >= System_Tab.N_MBOX)
        return(ERR_RANGE)
    else mbox_pt=&(mailbox_tab[mbox_id]);
    wait_sema(mbox_pt->send_sem);
    enqueue_data(mbox_pt->queue, data_pt);
    signal_sema(mbox_pt->recv_sem);
    return(DONE);
}
```

Receiving from a mailbox is achieved by waiting on the receiving semaphore and then retrieving the message as shown below

```
/*****************************************************************
              receive a message from a mailbox by waiting
 *****************************************************************/

data_unit_ptr_t recv_mailbox_wait( ushort mbox_id){

    mailboxptr_t mbox_pt;
    data_unit_ptr_t data_pt;

    if (mbox_id < 0 || mbox_id >= System_Tab.N_MBOX)
       return(ERR_RANGE)
    else mbox_pt=&(mailbox_tab[mbox_id]);
    wait_sema(mbox_pt->recv_sem);
    data_pt=deque_data_unit(mbox_pt->queue);
    signal_sema(mbox_pt->send_sem);
    return(DONE);
}
```

A real-time application may want to check a mailbox for a message and then
continue without blocking if there are no messages. DRTK provides a system call
recv_mbox_notwait for this purpose shown below. If there are no messages available
a NULL pointer is returned.

```
/*****************************************************************
              receive a message from a mailbox without waiting
 *****************************************************************/

data_unit_ptr_t recv_mailbox_notwait( ushort mbox_id){

    mailbox_ptr_t mbox_pt;
    data_unit_ptr_t data_pt;

    if (mbox_id < 0 || mbox_id >= System_Tab.N_MBOX)
       return(ERR_RANGE)
    else mbox_pt=&(mailbox_tab[mbox_id]);
    if(mbox_pt->recv_sem.value > 0) {
       wait_sema(mbox_pt->recv_sem);
       data_pt=deque_data_unit(mbox_pt->queue);
       signal_sema(mbox_pt->send_sem);
       return(data_pt);
    }
    return(NULL);
}
```

There is a need for the receive routine with a time-out in a real-time system. We
have the system call *recv_mbox_timeout* for this purpose which checks the mailbox
of the caller first and if there are not any messages, it is delayed for a certain number
of ticks and the mailbox is checked again.

```
/****************************************************************
            receive a message from a mailbox with timeout
 ****************************************************************/

data_unit_ptr_t recv_mailbox_timeout( ushort mbox_id){

    mailbox_ptr_t mbox_pt;
    data_unit_ptr_t data_pt;

    if (mbox_id < 0 || mbox_id >= System_Tab.N_MBOX)
       return(ERR_RANGE)
    else mbox_pt=&(mailbox_tab[mbox_id]);
    if(mbox_pt->recv_sem.value >0){
       wait_sema(mbox_pt->recv_sem);
       data_pt=deque_data_unit(mbox_pt->queue);
       return(data_pt);
    }
    else delay_task(current_tid, System_Tab.N_DELAY);
    if(mbox_pt->recv_sem.value >0){
       wait_sema(mbox_pt->recv_sem);
       data_pt=dequeue_data_unit(mbox_pt->queue);
       return(data_pt);
    }
    return(ERR_NO_MSG);
}
```

5.4.3 High Memory Management Using Pools

Higher level memory management in DRTK is achieved by the use of *data unit pools*.
A task that needs to transfer data to another one gets a free data unit from a pool,
insert its data to the data unit and passes the data unit to the mailbox of the receiving
task. The receiver consumes data in the message and typically returns the data unit
to the pool. This way, recycling of the memory space is done without exhausting the
memory. We first define the pool structure *pool_t* in the file "pool.h" which has a
semaphore to be checked if there is available data unit in the pool and an array of data
units. The pool table type (*pool_tab_t*) is an array of pools allocated at compile time,
inline with our strategy of statically allocation of storage for deterministic runtime
behavior.

```
/****************************************************************
                 data unit pool structure
 ****************************************************************/
// pool.h

typedef struct{
    int state;
    semaphore_t sem;
```

```
    unsigned int next;
    data_unit_t queue[System_Tab.N_POOL_MSG];
} pool_t;

typedef pool_t *pool_ptr_t;
typedef pool_t pool_tab_t[System_Tab.N_POOL];
```

A pool is allocated by searching for an unused pool in the pool table *pool_tab* in file "pool.c" as shown below. When such an entry is found, it is allocated and its write semaphore is initialized to the value specified in the system table and the read semaphore is specified to 0 value.

```
// pool.c
/******************************************************************
                    allocate a pool
******************************************************************/

int make_pool(){

    ushort i, sem_id;

    for(i=0; i<System_Tab.N_POOL; i++)
        if (pool_tab[i].state!=ALLOCATED){
            pool_tab[i].state=ALLOCATED;
            if(sem_id=make_sem(System_Tab.N_POOL_MSG)<0)
              return(ERR_NO_SPACE);
            pool_tab[i].sem=sem_id;
            return(i);
        }
    return(ERR_NO_SPACE);
}
```

The *get_data_unit* function is used to obtain a free data unit from the specified pool. If there are no available data units, the caller is blocked on the pool sending semaphore.

```
/******************************************************************
            get a data_unit from a pool
******************************************************************/

data_unit_ptr_t get_data_unit_(ushort pool_id) {

    pool_ptr_t pool_pt;
    data_unit_ptr_t data_pt;

    if (pool_id < 0 || pool_id >= System_Tab.N_POOL)
        return(ERR_RANGE)
    else pool_pt=&(pool_tab[pool_id]);
    wait_sema(pool_pt->sem);
    data_pt=dequeue_data_unit(pool_pt->queue);
```

```
        return(data_pt);
}
```

The *put_data_unit* function is used to place a used data unit to the specified pool.
If there is a waiting task on the pool semaphore for a free buffer, it is signalled.

```
/******************************************************************
                    put a data_unit to a pool
******************************************************************/

int put_data_unit( ushort pool_id, data_unit_ptr_t data_unit_pt){

    pool_ptr_t pool_pt;
    data_unit_ptr_t data_pt;

    if (pool_id < 0 || pool_id >= System_Tab.N_POOL)
        return(ERR_RANGE)
    else pool_pt=&(pool_tab[pool_id]);
    enqueue_data_unit(pool_pt->queue, data_pt);
    signal_sema(pool_pt->sem);
    return(DONE);
}
```

5.4.4 Task Management

The first thing that needs to be done is to create the task, initialize its variables,
assign *ready* state to it and insert it in the related queue by the *make_task* system
call. This function has an input parameter *sched* which is used to determine whether
the scheduler should be entered after making a task.

```
// task.c
/******************************************************************
                    make a task
******************************************************************/

int make_task(task_addr_t task_addr, ushort type, int priority,
ushort sched) {

    task_ptr_t task_pt, ushort i;
    for(i=0; i<System_Tab.N_TASK; i++)
      if (task_tab[i].state!=ALLOCATED){
        task_tab[i].state=ALLOCATED;
        task_pt=&(task_tab[i]);
        task_pt->address=task_addr;
        task_pt->type=type;
        task_pt->priority=priority;
        task_pt->state=READY;
        switch (task_pt->type) {
          case SYSTEM    : enqueue_task(&system_queue, task_pt);
                           break;
```

```
              case REALTIME   : insert_task(&realtime_queue, task_pt);
                                break;
              case USER       : enqueue_task(&user_queue, task_pt);
              }
              pthread_create(NULL, NULL, task_addr,NULL);
              if (sched) Schedule();
              return(i);
          }
      return(ERR_NO_SPACE);
}
```

Deleting a task removes it from the DRTK by freeing its task control block. The resources it holds must also be freed for the use of other tasks which we do not show for simplicity.

```
/*****************************************************************
                          delete a task
 *****************************************************************/
int delete_task( ushort task_id ){

    task_ptr_t task_pt;
    if (task_id < 0 || task_id >= System_Tab.N_TASK)
        return(ERR_RANGE);
    task_pt=&(task_tab[task_id]);
    task_pt->allocation=NOT_ALLOC;
    // free its resources
    return(DONE);
}
```

5.5   Initialization

We use a second way to access global data in the file "global_data.h" which is mainly for parameters needed by the application tasks whereas the system table contains data for system tasks in general. This file also includes all of the needed headers and declaration of all tables along with error codes, hence including this file only will provide all the necessary data structures visible to the application task.

```
// global_data.h
/*****************************************************************
                          global data
 *****************************************************************/
#define    DONE           1
#define    YES            1
#define    NO             0
#define    ERR_RANGE     -1    // range error
#define    ERR_NO_SPACE  -2    // no space in table
#define    ERR_RES_NOTAV -3    // resource not available
```

```
#define      ERR_NO_MSG    -4    // no messages in mailbox
#define      ERR_INIT      -5    // initialization error
#define      ERR_NOTAV     -6    // not available


#define      UNICAST       0x00 // message types in the network
#define      MULTICAST     0x10
#define      BROADCAST     0x20
system_tab_t System_Tab; // allocation of system table
task_control_block_t task_tab[system_tab.N_TASK]; // task table

task_queue_t system_queue;    // scheduling queues
task_queue_t realtime_queue;
task_queue_t user_queue;
task_queue_t delta_queue;      // delayed task queue

device_tab_t dev_tab;          // device table
semaphore_tab_t sem_tab;       // semaphore table
mailbox_tab_t mailbox_tab;     // mailbox table
pool_tab_t pool_tab;           // pool table

ushort current_tid;        // current task id
task_ptr_t  current_pt;    // current task pointer
```

The initialization routine *init_system* in file "init.c" allocates main data unit pools and makes system tasks and application tasks.

```
// init.c
/****************************************************************
                system initialization
****************************************************************/

int init_system(){
    int pool_id;
    int task_id;
    if(pool_id=make_pool() < 0) // make network pool
      return(ERR_INIT);
    System_Tab.network_pool_id=pool_id;
    if(pool_id=make_pool() < 0) // make user pool
      return(ERR_INIT);
    System_Tab.userpool1=pool_id; //

    if(task_id=make_task(DL_Out, SYSTEM, 0, NO)<0)
      return(ERR_INIT);
    System_Tab.DL_Out_id=task_id;
    /* make all other needed system tasks such as Timer_ISR,
       Clock_Synch, Leader_Elect etc. */
    if(task_id=make_task(Scheduler, SYSTEM, 0, YES)<0)
      return(ERR_INIT);
}
```

For simplicity, all of the header files are included in the main header file "drtk.h" as below

```
//drtk.h
/********************************************************
                    DRTK modules
 *******************************************************/
#include    "system_data.h"
#include    "global.h"
#include    "data_unit.h"
#include    "task.h"
#include    "isr.h"
#include    "device.h"
#include    "semaphore.h"
#include    "mailbox.h"
#include    "pool.h"
#include    "global_data.h"
```

Modules described until now are included as C source code in the "drtk.c" file as shown below. Any real-time application simple includes this file to have DRTK running.

```
/********************************************************
                    DRTK modules
 *******************************************************/
#include    "data_unit_queue.c"
#include    "task_queue.c"
#include    "schedule.c"
#include    "isr.c"
#include    "task_state.c"
#include    "device_man.c"
#include    "semaphore.c"
#include    "mailbox.c"
#include    "task.c"
#include    "pool.c"
#include    "init.c"
```

5.6 Testing DRTK

We have the DRTK simulation in a UNIX environment where each task is simulated by a POSIX thread. In order to have the simulation follow a real-life situation as much as possible, a POSIX thread started by the DRTK is made *ready* and put in a ready queue waiting to be started by the scheduler. The *Scheduler* itself is a POSIX thread that is always running and activated only by its semaphore. Also, we must have the tasks which are POSIX threads to execute according to the decision of the *Scheduler* instead of random execution of threads by the underlying UNIX operating system. In order to achieve this function, we have each thread initially wait on its

POSIX semaphore when activated. The *Scheduler* will signal the semaphore of a thread when it is scheduled. We have the producer/consumer example coding of which is shown below.

```c
// test.c
 #include <stdio.h>
 #include <pthread.h>
 #include <synch.h>
 #include "drtk.h"
 #include "drtk.c"

/****************************************************************
                        Producer
****************************************************************/

char c;
semaphore_t sem_prod,sem_cons;

TASK Producer(int *me){
    sem_wait(&(task_tab[me].sched_sem));

    do {
      c=getc();
      signal_sema(sem_cons);
      wait_sema(sem_prod);
    } while (c!=EOL)
}

/****************************************************************
                        Consumer
****************************************************************/

TASK Producer(void *me){
    sem_wait(&(task_tab[me].sched_sem));

   do {
      wait_sema(sem_cons);
      putc(c);
      signal_sema(sem_prod);
    } while (c!=EOL)
}

void main(){
   init_system();
   if((sem_prod=make_sema() < 0)||(sem_cons=make_sem() < 0))
      return(ERR_SYS);
   make_task(Producer, USER, 2, NO);
   make_task(Consumer, USER, 1, YES);
}
```

For separate compilation, we need to compile the DRTK code, the test code and then make an executable as follows. We then run the executable *test*.

```
gcc -c  drtk.c
gcc -c  test.c
gcc -o  test drtk.o test.o -lpthread
```

5.7 Review Questions

1. What are the main features of DRTK that makes it real-time?
2. What other fields may be needed in a task control block in a real application?
3. What is the function of the *state* field in a data structure in DRTK?
4. Why is a real POSIX semaphore needed for the correct scheduler functioning in DRTK?
5. What are the system calls for task synchronization in DRTK that may be needed by a real-time application?
6. What are the system calls for task communication in DRTK that may be needed by a real-time application?
7. Why system calls with time-outs are needed in DRTK?

5.8 Chapter Notes

We described the design and implementation of a distributed real-time kernel simulator in this chapter. The layering of DRTK is almost universal, similar to various operating system kernels, with the Scheduler at the bottom, task state management system calls activating the Scheduler when needed, task synchronization routines on semaphores calling task state management routines and finally the intertask communication routines to send/receive messages employing mailboxes by calling semaphore routines. The data structure design approach is similar to UNIX with each object identified by an integer that is its index in a statically allocated table. A device access is achieved through its device control block with global read and write operations that call the device drivers adddressed in this data structure, similar to UNIX internals.

However, system calls are implemented aiming at a real-time application. For example, the Scheduler is intended to be real-time with a real-time queue, and the semaphore and mailbox system calls may return without blocking and with time-out. Network communications and various middleware functions will be implemented on top of this kernel in the following chapters, to be able to invoke it in a distributed real-time environment.

5.9 Programming Projects

1. The *enqueue* and *dequeue* operations on data units and task control blocks are similar. Show how to implement a single *enqueue* and a single *dequeue* function for both data types.

Table 5.1 An example task set

τ_i	C_i	T_i
1	3	18
2	8	36
3	5	24

Fig. 5.6 Task graph for Exercise 4

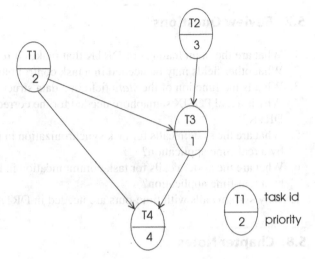

2. Write a test program that is a producer consumer which exchange data by messages using mailboxes.

3. Three independent periodic hard real-time tasks with the characteristics shown in Table 5.1 are to be implemented using DRTK. Write the application program in C by using all needed DRTK system calls.

4. Four dependent aperiodic real-time tasks are shown in the task graph of Fig. 5.6. Task T_1 and T_2 gets 20 characters from the user, T_1 sends the first 10 characters to task T_3 and the second 10 characters to task T_4. Task T_2 sends all 20 characters to task T_4 which prints all data it receives. Form all these tasks and the main program in C language which are implemented using DRTK where the basic communication is by mailboxes.

References

1. Comer D (2015) Operating system design: the XINU approach, 2nd edn. Chapman and Hall/CRC
2. Erciyes K (1989) Design and realization of a real-time multitasking kernel for a distributed operating system. PhD thesis, Ege University

Distributed Real-Time Operating Systems and the Middleware

6.1 Introduction

A fundamental requirement from a distributed operating system (DOS) is the management of resources over the network, with the network itself being one such resource. We need to add the word *timely* when a distributed real-time operating system (DRTOS) is considered. The collapsed OSI model for a distributed real-time system (DRTS) typically consists of the data link layer and the physical layer and possibly a thin transport layer only. A DRTOS should provide an efficient interface to data link layer functions in the medium access control and logical link control sublayers of the data link layer. In doing so, the DRTOS should enable the real-time characteristics of the network to be reflected in the application. For example, when a real-time application task sends a message with time-out to another task residing on a different node of the network, this function should be translated to the related data link function that performs the required operation in real-time. In many applications in a DRTS, only a subset of tasks is involved cooperating to finish a common job. This operation requires multicast communications, and *task* (or *process*) *groups* implementation is an effective method to achieve this form of communication. Additionally, task groups are commonly used for task replication for fault tolerance. All copies of a task are included in a group and they execute the same code. If the primary task fails, a copy is selected to run as primary. However, the order of messages sent to all copies should be the same to have a consistent state in the group and realizing this function requires detailed analysis. Task group management is typically handled by the DRTOS or sometimes by the middleware. The processors are the main resource to be managed by the DRTOS. The static scheduling of tasks to processor can be achieved offline and dynamic load balancing requires fair distribution of the load at runtime. We leave the review of these methods to the next part when scheduling in a DRTS is studied in detail.

© Springer Nature Switzerland AG 2019

K. Erciyes, *Distributed Real-Time Systems*, Computer Communications and Networks, https://doi.org/10.1007/978-3-030-22570-4_6

The *middleware* is the set of software modules that lie between the application and the operating system. This definition encompasses single computational nodes and more commonly, distributed systems. The application typically requests a service from the middleware and the middleware calls an operating system function to realize the request. A fundamental reason for implementing such a software layer is to provide re-usability. Many diverse applications require similar functions which are not contained in common operating system tasks such as processor, memory, input/output, and process management. Design and implementation of these tasks for each application are unnecessary in many cases. The real-time middleware works similar to a general middleware software with the addition of performing the required functions in real time. Some middleware modules may be explicitly needed in a DRTS. In this chapter, we review typical middleware modules that are needed in most DRTSs; clock synchronization among the nodes of a DRTS, task groups, and leader election. We conclude the chapter with the implementation of the network interface, task groups, clock synchronization and leader election in the sample distributed real-time kernel (DRTK).

6.2 Distributed Real-Time Operating Systems

The collapsed model of the OSI 7-layer model typically has the data link layer, the physical layer, and a transport layer interface to the application as stated. Note that we can discard the network layer assuming the real-time network is localized without the need for significant routing procedures. An important requirement from the DRTOS is to manage the datalink layer functions and interface them to the operating system. However, the application software does not and should not, in general, be aware of data link functions which means there is a need for a higher level above the data link layer to provide access point interface to the DRTOS. This layer typically has the characteristics of the transport layer of the OSI model. The modified OSI model for a DRTS then can be considered to have the layers as shown in Fig. 6.1. Here, we can see that the actual real-time network interface at physical layer is handled by the device drivers and the DRTOS considers the network as a device. Note that data link layer functions to be realized by the DRTOS and the protocol may overlap. The main functions to be performed by a DRTOS are the management of data link layer and the interface between the application tasks and the data link layer functions as described in the next sections.

6.2.1 Transport Layer Interface

The main service performed at transport layer is directly connecting and communicating to an application on another host computer. Note that any layer below (network and data link) does not differentiate among the application messages. Two application tasks are logically connected through this layer enabling end-to-end application

Fig. 6.1 Collapsed OSI model used in a real-time network

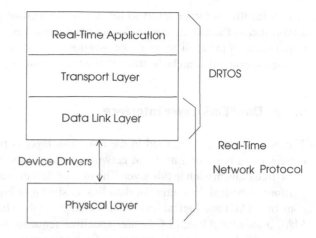

connection and communication. For example, a web client at a host communicates directly to another web server in another host using the transport layer. Since a host computer will have many applications running, an important function of transport layer is to provide multiplexing and demultiplexing among the application tasks. In order to do so, the application layer message is divided into smaller protocol data units (PDUs), transport layer header is added, and then the message is passed to network layer as a Layer 4 PDU. The transport layer at the receiving host removes the PDU header, reassembles the message, and passes it to the application. The two main Internet transport protocols are transmission control protocol (TCP) which provides reliable, connection-oriented data transfer with flow control and the user datagram protocol (UDP) which is a connectionless service without any flow control.

Based on these general functions realized at this layer, we will consider possible adaptation for the DRTS case. We will assume the nodes in the DRTS are physically located close to each other as in most of the DRTS applications. We can, therefore, assume the network layer is almost nonexistent and can make a direct connection to the data link layer. The main task for the DRTOS at transport layer is then dividing the application message into PDUs and reassembling them at the receiving side. Accordingly, two basic routines that are APIs for the real-time tasks at this layer are as follows.

- *send_message(message, task-id)*: Send a message to another task that may reside on another host.
- *receive_message(message)*: Receive message from any task.

We need to consider the case of real-time tasks, which means we need variations of the *send* routine such as *send with time-out* and accordingly, *receive with time-out*. The *send* routine should also distinguish between local and remote tasks in order to pass the message blocks locally if the receiving task is on the same host as the sender. This function can be realized by a *naming* mechanism. In the simplest form, a range of task identifiers can be reserved for the local tasks, and any identifier larger than a limit means the task is remote. Yet another critical issue is to pass the timeliness requirement of the application all the way down to the network protocol. We will see how these problems can be handled in the next sections.

6.2.2 Data Link Layer Interface

The general function required in the data link layer is point-to-point flow control between two hosts or a host and network device in the network, and provision of error-free transmission in this layer. The unit of data exchanged at this layer is called a *frame*. A typical data frame of data link is shown in Fig. 6.2 with medium access control (MAC) and logical link control (LLC) fields. The *MAC header* is used for MAC address and flag; *LLC header* specifies sequence number and the type of the frame; and *data* is the basic datagram unit communicated with the network layer. The MAC trailer is used for error detection and correction with cyclic redundancy check (CRC) being one the most commonly used error detection methods.

Assuming we have a real-time network protocol such as CAN, TTP, or IEEE 802.11e, we need an interface between the data link drivers of such a protocol and the DRTOS. As in our DRTK realization, we can consider the network as any other device in the operating system. The device control block data structure associated with the real-time network (network control block) can now contain the activation code for all drivers needed to perform the required mode of communication in the network. The next step to implement the interface is to form basic data link layer functions as viewed by the DRTOS. A convenient way to achieve this goal is to have the below real-time system tasks specified at data link layer interface. These tasks communicate with the upper transport layer using their mailboxes, which are data structures to deposit messages asynchronously (see Sect. 4.3.3).

- *DL_Output*: This task continuously waits at its mailbox, and when there is a message to be transferred, it receives the message and activates the network driver using the network control block. When the application task needs to wait for a specific time for a reply, this task sleeps for a predetermined time and then wakes up to see if any response is received.
- *DL_Input*: This task reads a frame from the network using the network control block driver and checks for errors. In case of an error, it transmits a negative acknowledgement to the sender. Note that this way, we are implementing error checking and flow control. When the network protocol provides this functionality,

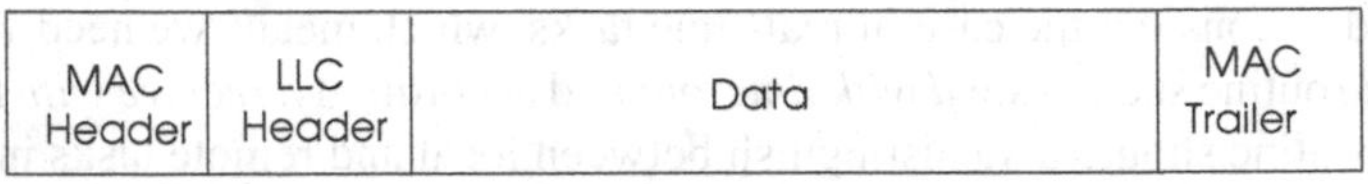

MAC Header	LLC Header	Data	MAC Trailer

Fig. 6.2 Data link layer frame

the job of this task are simply read a message from the network and send it to the mailbox of the receiver. In the case of a *receive with time-out*, this task reads from the network by activating the input driver and if there are no messages, it sleeps for a predetermined amount of time.

Figure 6.3 illustrates how these tasks at data link layer interface are related and the way transport layer is attached to them. When the *DL_In* task discovers the message

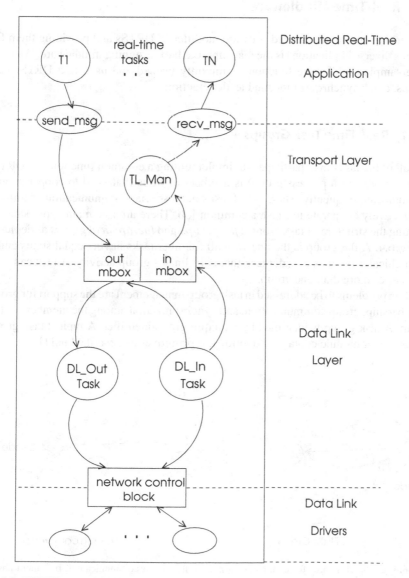

Fig. 6.3 DRTOS network interface, modified from [3]

is destined to another node, it simply deposits the message in this mailbox for the *DL_Out* task to take it from there and send it to the network. Using this mailbox for this purpose may be practical so that no data link header and error check is needed in this case. The transport layer manager (*TL_Man*) is a task that is used to switch the incoming application messages to the recipient tasks.

6.3   Real-Time Middleware

Diagnosing commonly needed software modules in DRTSs and providing them for various diverse applications is the main approach of designing middleware. We have three sample middleware functions commonly needed and used in a DRTS; task groups, clock synchronization, and leader election.

6.3.1   Real-Time Task Groups

Not all of the tasks may participate in implementing a common function in a DRTS. A *task group* (or a process group) is a subset of tasks that need to cooperate and communicate frequently. These set of tasks need special communication primitives to effectively cooperate to finish a common job. There are two main approaches in forming the structure of task groups; *flat groups* and *hierarchical groups* as depicted in Fig. 6.4. A flat group is the one with all member tasks having equal status and a hierarchical group has a leader task that coordinates group activity. A task can be a member of more than one group.

Main problems to be addressed in task group management are the support for group membership, group communication, and synchronization among the members of the group. A task group is identified by a unique group identifier. A typical task group management module contains the following primitives as described in [1]:

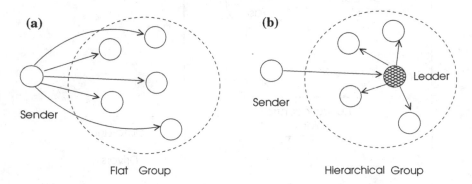

Fig. 6.4 Group structures for communication. **a** A flat group communication. **b** A hierarchical group communication

- *group-id = CreateGroup(initial task id, type)*: The calling task creates a group with a unique identifier and becomes a member, and leader of the group if it is a hierarchical one.
- *JoinGroup(group-id, task-id)*: Calling task becomes the member of the specified group based on access rights.
- *LeaveGroup(group-id, task-id)*: Calling task removes itself from the group. If it is the leader of a hierarchical group, a new leader is elected.

Sending a message to a group can be performed using different methods based on the structure of the group. Employing an elected group leader eases the implementation of group communications and other group management routines. If the group is a flat one, the message has to be delivered to all members of the group by the transport layer routine. The receiving members may wait at their local mailboxes or at a group mailbox specifically used for multicast communications. For the hierarchical group, the message is delivered to the leader which deposits it in mailboxes of all of the members or in the group mailbox. These operations are depicted in Fig. 6.4. Basic group communication procedures are as follows:

- *SendMcast(message, group-id)*: The message is sent to all members of the group. This can be achieved by sending the message to the group leader which, in turn, sends the message to all members of the group as stated.
- *RecvMcast(group-id, task-id)*: Receive a multicast message either from the group mailbox or from the leader, or from the private mailbox.

Task groups are commonly employed for redundancy in fault-tolerant DRTSs. Copies of a critical hard real-time task are formed as a group and these tasks execute in parallel with the foreground (primary) task. If the primary task fails, one of the secondary (background) tasks is selected to run as the primary. The operation of this fault tolerance requires careful consideration of how to start a secondary task when a failure occurs. The provision of reception of all multicast messages in the same order is crucial for correct operation as we will see in Chap. 12. Considering the case when all secondaries are running the same FSM will lead to different states of the copies when messages are received in different orders making the recovery from fault erroneous. The parallel processing programming tool Message Passing Interface (MPI) provides a convenient interface for group management and communication to the application [11].

6.3.2   Clock Synchronization

Agreement on a common time frame is crucial in a DRTS as in a non-real-time distributed system for various reasons. Failure to do so may result in erronous states of the nodes which may lead to collapse of the whole system. For example, missing deadlines of tasks distributed over the network due to different clock values at nodes becomes inevitable. In a DRTS, the usual requirement is to have the clocks synchronized to an external real clock or a reference clock within allowed margins. The universal coordinated time (UTC) keeps time based on the atomic clock with corrections needed because of the variations in the rotation of the earth. The UTC clock value can be accessed via satellites which propagate it constantly. The global positioning system (GPS) allows to determine the geographical position of an entity [14]. A GPS satellite broadcasts its position with a timestamp allowing accurate reception of time by receivers.

Each computing element of a DRTS is equipped with a hardware clock which is fed by a crystal. Although frequency of each crystal is stable, frequencies of crystals at nodes may have different frequencies which cause the clock values drift away from each other in time. Keeping the time at a node is performed by having a timer that decrements its value at a certain rate per second and when this value reaches zero, timer interrupt is generated. The timer interrupt handler then increments the clock value of the node to reflect the current time. The clock synchronization algorithms attempt to correct the drifting clock values of nodes in the DRTS. Two cases may be considered for this purpose; a DRTS with one or more nodes that keep their time with GPS and all other nodes synchronized to these nodes, or none of the nodes have GPS receivers and they are to be synchronized to each other within allowable limits.

6.3.2.1 Logical Clocks

Instead of synchronizing physical clocks, ordering of events can be provided by *logical clocks*. In this case, our main concern in terms of synchronization of clocks is the ordering of events that happen at various nodes of the distributed system. The *happened before relation* ($\rightarrow$) can be used to describe the casual ordering of events with the following rules:

- If a and b are events at the same task of a node of the distributed system, and a happens before b, then $a \rightarrow b$.
- If a is the event of sending of a message m by some task and b is the receipt of this message by some other task, then $a \rightarrow b$.
- If $a \rightarrow b$ and $b \rightarrow c$, then $a \rightarrow c$.

Two events a and b are said to be concurrent ($\|$) if neither $a \rightarrow b$ nor $b \rightarrow a$. holds. We can implement these rules to synchronize logical clocks in a distributed real-time system with the following assumptions:

- A task τ_i has a logical clock C_i.
- A clock C_i assigns a value of $C_i(a)$ to an event a that happens at task τ_i.
- The values assigned by the clock C_i monotonically increase.

In order to provide synchronization of the logical clocks, the following conditions must be satisfied:

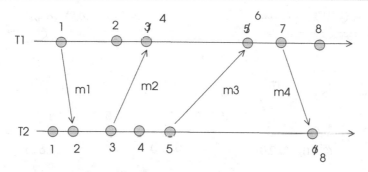

Fig. 6.5 Logical clock synchronization between two tasks

- If two events a and b occur at the same task τ_i with a preceding b, then $C_i(a) < C_i(b)$.
- If a is the event of sending of a message m in task τ_i and b is the receipt of this message in task τ_j, then $C_i(a) < C_j(b)$.

Lastly, in order to enforce the above conditions to hold, the following implementation rules must be obeyed for time synchronization using logical clocks:

1. Clock C_i is incremented between any two consecutive events at task τ_i.
2. a. If a is the sending of a message m in task τ_i, then m is timestamped with $t_m = C_i(a)$.
   b. If b is the event of receiving the message m in task τ_j, then the clock C_j of task τ_j is set as follows:

$$C_j \leftarrow max(C_j, t_m + 1)$$

Synchronization of logical clocks between two tasks τ_1 and τ_2 is depicted in Fig. 6.5. The clock values are shown to increase monotonically and when τ_2 receives message m_1, there is no change in its clock value. However, messages m_2, m_3, and m_4 cause changes in local clock values when the above rules are implemented.

6.3.2.2 Vector Clocks

The main problem with the logical clocks is that if $C(a) < C(b)$, we cannot determine that $a \to b$. *Vector clocks* are an improvement over the logical clocks by employing a clock vector $C_i[1 \dots n]$ at each task τ_i, where $C_i[i]$ is the logical clock value of task τ_i and $C_i[j]$ is the estimation of the clock of τ_j by τ_i. We now have the following implementation rules for vector clocks:

1. Clock C_i is incremented between any two consecutive events at task τ_i.

$$C_i[i] \leftarrow C_i[i] + 1$$

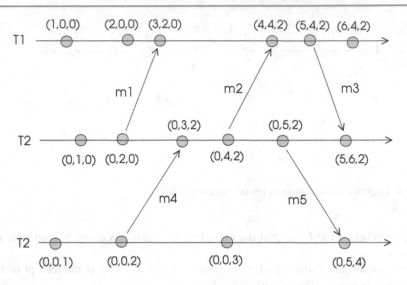

Fig. 6.6 Vector clock synchronization among three tasks

2. a. If a is the sending of a message m in task τ_i, then m is timestamped with $t_m = C_i(a)$ where t_m is the vector clock stored in the message.
    b. If b is the event of receiving the message m in task τ_j, then the clock C_j of task τ_j is set as follows:

$$\forall k, C_j[k] \leftarrow max(C_j[k], t_m[k])$$

Two vector timestamps $t(a)$ and $t(b)$ for two events a and b are equal if and only if $\forall i, t^a[i] = t^b[i]$. The less than or equal relationship between $t(a)$ and $t(b)$ means, $t^a \leq t^b$ if and only if $\forall i, t^a[i] \leq t^b[i]$. With the above rules, we can say we can determine $a \rightarrow b$ if and only if $t_a < t_b$ which is a solution to the problem of logical clocks. Three tasks in Fig. 6.6 are synchronized using vector clocks using these implementation rules.

6.3.2.3 Network Time Protocol

The network time protocol (NTP) assumes the existence of a server in a distributed system that receives accurate clock values by the UTC broadcasts [9]. Clock synchronization is achieved by the nodes acquiring the time from the server at regular intervals. However, delays introduced by the network have to be taken into account when correcting the node clock values. The timestamp used by NTP is 64 bits; 32-bit for seconds and a 32-bit for the fractional second. The latest NPv4 doubled the length of these fields to give a timestap of 128 bits [10].

The two parameters needed for synchronization are the *time offset* and *round-trip delay*. Consider the case when a node is attempting to correct its clock by sending a message timestamped at t_1 to the server as shown in Fig. 6.7. This message is

Fig. 6.7 NTP example. Time values are in milliseconds

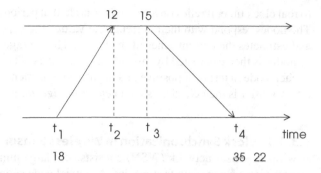

received by the server at time t_2, timestamped with this value, responded at time t_3 which is again timestamped and the sending node receives the response at time t_4.

The sending node has all of these time values and can compute the time offset Θ as below.

$$\Theta = \frac{(t_2 - t_1) + (t_4 - t_3)}{2} \tag{6.1}$$

The round-trip delay δ observed by the sender is the difference between the total time passed at sender and processing at the server as follows:

$$\delta = (t_4 - t_1) - (t_3 - t_2) \tag{6.2}$$

It is assumed that the propagation delay in the network in the reverse direction is the same, in other words, the network is symmetric. After statistically analyzing the obtained values, the lowest value is selected and the local node time is corrected using this value. The instant Θ and δ values, for example, in Fig. 6.7 are 7 and 14, respectively. The receiver would add half of the round-trip delay to the clock value of the server, correcting its clock to 22. When a group of servers have more accurate clocks than others, they are given a higher priority using a classification scheme. A less complicated version of NTP called the simple network time protocol (SNTP) is used in some embedded applications [13].

6.3.2.4 The Berkeley Algorithm

The Cristian's clock synchronization algorithm assumes the existence of an accurate time server that receives time values from UTC [2]. A node requests time from the server, adds the half of the round-trip time (RTT) to the timestamp in the reply message to determine its clock value. It assumes the delays are equal in both directions over the network which is reasonable for short RTTs.

The Berkeley algorithm uses a similar estimation of RTT and is designed mainly for Intranets. It works with no UTC information and hence, there is no external clock value reading and adjustment in this algorithm, easing the implementation. Basically, relative synchronization among nodes is aimed rather than synchronizing to real clock time. It relies on a *supervisor* node that periodically polls *worker* nodes. The nodes respond with their current time value. The supervisor considers the RTT and estimates the current times of the nodes. The average value of the clock values of nodes is then calculated by discarding the outliers. The correction values are sent to each node in terms of positive and negative adjustments. The master may fail and a new master is then elected using an *election algorithm*.

6.3.2.5 Clock Synchronization in Wireless Sensor Networks

A wireless sensor network (WSN) consists of a large number of sensor nodes that communicate using radio frequencies. A central node called the *sink* has more computational power and facilities than the ordinary nodes in a WSN. A sink is used to gather data from the nodes, analyze and possibly relay this information to a remote computer for finer analysis. A node in a WSN has a limited lifetime as its basic source of energy is its battery. Moreover, communication over the wireless medium dissipates significant energy, requiring different and simpler clock synchronization algorithms in these networks. We will review two such protocols: The Reference Broadcast Synchronization Protocol and the Timing-Synch protocol.

Reference Broadcast Synchronization Protocol

The Reference Broadcast Synchronization (RBS) protocol is different than the traditional synchronization protocols as the receivers are synchronized with each other rather than a sender and a receiver getting synchronized [4]. In its simplest form, a message *m* with no timing information is broadcast by a WSN node say, to two receivers. Each receiver records the time it receives the message *m*, and then exchange this information with each other to know the time difference between them and provide the necessary correction. The main advantage of the RBS is the removal of the sender from any offset calculation. Propagation delay and receiver times cause the offset and the former can be neglected assuming the broadcast message is received by all the nodes of the WSN simultaneously when the number of hops between two farthest nodes, that is, the diameter of the network graph is small.

Time-Synch Protocol

The Timing-Synch Protocol for sensor networks (TPSN) considers the time of the sender and the time of the receiver during synchronization [6, 12]. It has two phases: the level discovery phase and the synchronization phase. The first phase is run once when the network is deployed. The sink node starts the algorithm by sending a broadcasts *level_discovery* packet which contains 0 as its level and its identifier to its neighbors. Any node receiving this message for the first time assigns its level to one plus the level in the message, inserts this value and its identifier in the message and then broadcasts. This way, a spanning tree rooted at the sink (root) node is constructed. Algorithm 6.1 displays the possible implementation code for the level discovery phase.

Algorithm 6.1 *Level Discovery Algorithm*

```
1:
2: procedure LEVEL_DISCOVERY
3:    int my_id, my_parent, my_level = 0, flag = 0;
4:    message msg;
5:    if my_id = sink then
6:       my_parent ← my_id
7:       msg.id ← my_id
8:       msg.level ← 0
9:       send broadcast msg
10:   else
11:      receive msg;
12:      if flag = 0 then
13:         my_level ← msg.level + 1
14:         my_parent ← msg.id
15:         msg.id ← my_id
16:         msg.level ← my_level
17:         flag ← 1
18:         send broadcast msg
19:      end if
20:   end if
21: end procedure
22:
```

The synchronization phase is performed periodically by the root by broadcasting a synchronization message $time\_synch$. Each level 1 node that receives this message starts a two-way message exchange with the root after waiting for a random amount of time to avoid collisions. A level 2 node that overhears level 1 nodes communicating with the root then initiates a two-way communication with a level 1 node and so on until all nodes at all levels are synchronized. The two-way communication occurs as in Fig. 6.7 with shown times t_1, t_2, t_3 and t_4. The initiating node A at level i inserts its level, timestamps it and sends it to the node B at level $i - 1$ at time t_1. The node B receives this message at time $t_2 = t_1 + \Delta + d$ where Δ is the relative clock drift between the nodes, and d is the propagation delay of the message. The response from node B sent at time t_3 contains its level number and times t_1, t_2 and t_3 and this message is received by A at time t_4. Node A can now calculate Δ and d as follows and adjusts its clock according to node B.

$$\Delta = \frac{(t_2 - t_1) - (t_4 - t_3)}{2}; \quad d = \frac{(t_2 - t_1) + (t_4 - t_3)}{2} \tag{6.3}$$

Assuming propagation delays are constant in each direction, the clock drift between the two nodes and the propagation delay can be accurately determined. It is claimed that TPSN achieves two times better precision than RBS [6]. The role of the root node may be taken over by other sensor nodes in turn by a leader election algorithm, when there is no such node in the WSN.

6.3.3   Election Algorithms

Various middleware functions in a DRTS can be realized simpler and more effectively when some of the main functions in a group of tasks/nodes are handled by a *leader*. For example, grouping a number of nodes in a cluster and electing one of them as the leader of the cluster is an efficient method used for routing in wireless sensor networks. We saw also how a coordinator/leader of a group may be used for sending group messages. The leader of a group may fail and there is a need for electing a new leader in this case. *Leader election* algorithms address this problem as described in the next section.

6.3.3.1   Election in a Unidirectional Ring

We will consider an election algorithm using an FSM to be implemented in a unidirectional ring with each node having a unique identifier. The general idea of this algorithm is that when an ordinary node detects the leader is down, it starts an election among the running nodes by an *election* message. It inserts its identifier in the election message and changes its state to ELECT. Any node that receives this message $m(j)$ checks the content and does one of the following, with i being the identifier of the receiver.

- $i > j$: Node i replaces j with i in message and passes it to the next node. It also changes its state to ELECT.
- $i < j$: Node i does not alter message contents, it passes the message to next node and changes its state to ELECT.
- $i = j$: The *election* message has made one full traversal of the ring and node i has the highest identifier among running tasks and thus becomes the leader. It sends the *leader* message to its next neighbor.

This way, the highest identifier task is ensured to be the leader. The states of nodes are shown in Fig. 6.8. Note that only the leader can be in LDR state while any other node is in IDLE state when the leader is present.

Another leader election algorithm is the *Bully Algorithm* in which a node P_i detecting the failure of the leader sends an election message to all higher identifiers than itself. If there is no response in a predetermined time interval, P_i becomes the leader and sends a *leader* message to all nodes declaring itself the leader. Any node that receives a *election* message and has an identifier greater than that of the sender, sends back a reply message and it starts the election itself. A node that receives a reply to its election message is out of the election process and waits for the leader message to determine its leader.

6.3.3.2   Election in Wireless Sensor and Mobile Ad Hoc Networks

A WSN is a large-scale DRTS operating with many hard real-time tasks. Leaders are commonly needed in a WSN mainly for routing purposes. A WSN can be partitioned

Fig. 6.8 FSM of the leader
election algorithm in a ring

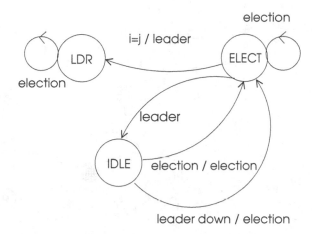

to a number of group of nodes called *clusters*. Using clusters, topology management of a WSN is simpler, this method also minimizes the size of the routing tables stored at the individual nodes. A broadcast message may be relayed to the leader of a cluster of WSN nodes and the leader will transfer the message to the nodes in its cluster. Also, leaders can form the anchor points with GPS receivers when other nodes use them to find their geographical positions and synchronize their clocks. The leader (or the cluster head (CH)) of a cluster is mainly used for this purposes and when its energy level falls under a predetermined value, a new leader is elected. A mobile ad hoc network (MANET) consists of mobile nodes with wireless communication facilities. Multihop communications are the basic means of message exchange in these networks. As in a WSN, clustering and selecting a CH for each cluster provides a simple and effective method for routing. A MANET that is partitioned into four clusters; C_1, C_2, C_3, and C_4 is shown in Fig. 6.9 where node a sends a message to node b using the CHs. Node a knows only its CH and sends the message to it which broadcasts it to the higher level connection among the CHs. Each CH knows the identifiers of the nodes in its cluster and if the destination address is in its cluster, it relays the message to the destination node and stops broadcasting it to other CHs. This way, the size of the routing tables at each node and CH is kept minimum.

A CH in a WSN is selected based on criteria such as the energy level, connectivity, and mobility. Two types of CH election protocols in WSNs are generally studied; random algorithms and minimum finding algorithms. The Low-Energy Adaptive Clustering Hierarchy (LEACH) is a clustering algorithm for WSNs [7]. Each node in LEACH selects a random number between 0 and 1 and if this number is less than a threshold, that node becomes the CH for the current round. The elected CH broadcasts its election to cluster members. The main role of the CH in this algorithm is to construct a time-division multiple access (TDMA) for members to use the network and also inform them when to transmit. The CH also aggregates data from the members and uploads this data to the sink. LEACH-C considers the location of the nodes and their energy levels when selecting a CH [8]. HEED is a hierarchical

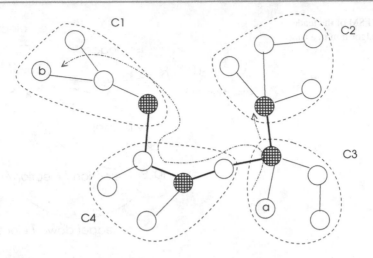

Fig. 6.9 Cluster-based routing in a MANET. CHs are shown in stripes

clustering method that selects CHs based on the residual energy of nodes and intra-cluster communication costs [5].

6.4   DRTK Implementation

We will describe how to implement the transport layer and data link layer interfaces and middleware modules of clock synchronization, group management and leader election in DRTK. We will assume that the stations are connected as unidirectional ring and messages are transmitted by forwarding them from a station to its successor. The sending node discards the message when it receives its own message. In our implementation, we will assume that only the physical layer but not the data link layer of the communication protocol exists.

6.4.1   Initializing the Network

We will assume three nodes of the network are connected in a unidirectional ring structure as shown in Fig. 6.10. Each node of the network is a UNIX process and can be coded as a separate program. The network communication between the nodes is achieved by the UNIX message queues.

We will call the nodes of the distributed real-time system as *node1*, *node2*, and *node3*. The code for *node1*, for example, will include "drtk.h" and "drtk.c", the usual initialization of DRTK and an additional network initialization routine *init_network* that creates a message queue creation routine is shown below in the file "network.c". We need to form the network device control block, form device drivers and store

Fig. 6.10 DRTK network structure

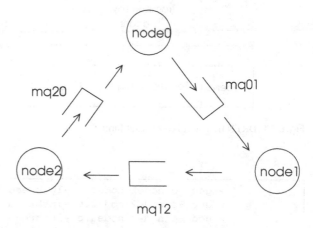

their addresses in this data structure so that data link tasks can call them. We omit error handling routines for simplicity.

```
// network.c
 #include <stdio.h>
 #include <pthread.h>
 #include <synch.h>
 #include <drtk.h>
 #include <drtk.c>

 long int keys[3]={1234L,2345L,3456L}, my_key_in, my_key_out;
 int msgq_id;

 int read_net(int *address, int n_bytes) {
   msgrecv(address, n_bytes, 0, IPC_WAIT);
 }

 int write_net(int *address, int n_bytes) {
   msgsnd(msgq_id_out, address, n_bytes, IPC_NOWAIT);
 }

 void init_network() {
   my_key_in=keys[Sys_Tab.this_node];
   msgq_id_in=msgget(my_key_in, IPC_CREAT | 0660);
   my_key_out=keys[(Sys_Tab.this_node+1) MOD 3];
   msgq_id_out=msgget(my_key_out, IPC_CREAT | 0660);
   net_dev=make_dev();
   net_dev.read_addr=read_net;
   net_dev.write_addr=write_net;
 }
```

We will need to code three programs with different node identifiers in the system table (*System_Tab.this_node*) and compile each node program individually, and then run these programs as shown below.

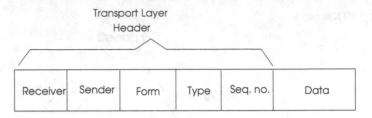

Fig. 6.11 DRTK transport layer packet format

```
> gcc -o node1 node1.c -lpthread
> gcc -o node2 node2.c -lpthread
> gcc -o node3 node3.c -lpthread
> node1 &
> node2 &
> node3 &
```

6.4.2 Transport Layer Interface

We will first specify the transport layer (TL) message structure as depicted in Fig. 6.11. This layer is the main interface to the application and our basic requirement is that the application is presented with a global task identifier without knowing where that task is. Thus, it is the job of this level to find whether the data transfer is local or remote or sometimes both in case of a multicast message. The *form* field specifies whether the message is unicast, multicast, or broadcast and the *type* field is a more detailed type specific to the application.

The TL header is specified in the file "data_unit.h" as below. For any application/middleware specific information, we will parse the data field of the PDU using the C *union* data type declaration.

```
// data_unit.h
/*************************************************************
                  Transport Layer Header
*************************************************************/
typedef struct {
    ushort sender_id;
    ushort receiver_id;
    ushort form;
    ushort type;
    ushort seq_num;
}TL_header_t;
```

The basic system calls at this layer are the *send_not_wait* and *recv_wait* routines to send and receive messages. These routines are used for both unicast and multicast message transfers. An important issue to be handled is to resolve whether the recipient(s) are local and/or remote. We adopt a simple naming scheme here, each task has a global 16-bit unique identifier which is formed by combining the unique host identifier with the local task identifier. For example, 0×1206 is the local task with *tid* = 06 on machine 18. We will assume the sender of the message knows the global identifier of the receiver task. The *send* routine should disassemble the message into a number of fixed size packets if needed, as shown in the code below in the file "transport.c". It gets a free data unit from its pool, fills it with message contents and deposits this data in the mailbox of the receiver using a non-waiting send. If the call is a multicast send, it searches for the local group identifier and deposits the message in the local group mailbox as well as storing it in the mailbox of the data link output task as below. We assume a task can be a member of one group only, identifier of which is stored in its control block. Similarly, a broadcast message is delivered to all active tasks locally and sent as a broadcast message over the network. Note that we do not have a separate multicast *send* but the application specifies the type of message as unicast, multicast or broadcast. The *send_net* routine prepares MAC layer header for the network and places the data unit at the mailbox of the data link output task.

```
/*  transport.c  */
/*****************************************************************
              send message without waiting
*****************************************************************/
int send_net(data_ptr_t data_pt, ushort tid, ushort f) {

    ushort mbox_id=task_tab[System_Tab.DL_Out_id].mailbox_id;
    data_pt->TL_header.form=f;
    data_pt->TL_header.sender_id=current_tid;
    data_pt->TL_header.receiver_id=tid & 0x00FF
    data_pt->MAC_header.sender_id=System_Tab.this_node;
    data_pt->MAC_header.receiver_id=(tid & 0xFF00)>>8;
    send_mailbox_notwait(mbox_id,data_pt);
}

int send_msg_notwait(ushort tid, char* msg_pt, ushort len,
                ushort type) {

    task_ptr_t task_pt;
    mailbox_ptr_t mbox_pt;
    data_unit_ptr_t data_pt1, data_pt2;
    ushort node_id, mbox_id, group_id, tid;

    if (tid < 0 || tid >= System_Tab.N_TASK)
        return(ERR_RANGE);
    if (type == UNICAST) {  // send unicast
      node_id=(tid & 0xFF00)>>8;
      if (node_id == System_Tab.this_node) //check remote
        mbox_id==task_tab[tid & 0x00FF].mailbox_id;
      while(len>0) {//do for local and remote
        data_pt1=get_data_unit(System_Tab.Net_Pool);
        memcpy(data_pt1, &(msg_pt.data), N_DATA_UNIT);
        if(node_id == System_Tab.this_node)
          send_mailbox_notwait(mbox_id, data_pt1);
        else
          send_net(data_pt1, tid, UNICAST);
        len=len-System_Tab.N_DATA_UNIT;
        msg_pt=msg_pt+len;
      }
    }
    else { // MULTICAST or BROADCAST
      while(len>0) {
```

```
           data_pt1=get_data_unit(System_Tab.Net_Pool);
           data_pt2=get_data_unit(System_Tab.Net_Pool);
       memcpy(&(msg_pt.data), data_pt1, System_Tab.N_DATA_UNIT);
       memcpy(&(msg_pt.data), data_pt2, System_Tab.N_DATA_UNIT);
       if (type == MULTICAST) {
          group_id=task_tab[current_tid].group_id;
          mbox_id=group_tab[group_id].mailbox_id;
          send_mailbox_notwait(mbox_id,data_pt1);
       }
       if (type == BROADCAST)
          for(i=0;i<System_Tab.N_TASK;i++)
             if (task_tab[i].state=ALLOCATED) {
                data_pt=get_data_unit(System_Tab.Net_Pool);
                memcpy(data_pt, &(msg_pt.data), N_DATA_UNIT);
                send_mailbox_notwait(task_tab[i].mailbox_id,data_pt);
             }
          send_net(data_pt2, tid, type);
          len=len-System_Tab.N_DATA_UNIT;
          msgpt=msgpt+len;
       }
    }
    return(DONE);
}
```

We have a transport layer manager task (*TL_Man*) that receives a message from
the data link layer at its input mailbox. It checks the transport layer header type to
decide whether the incoming message is a unicast, multicast or broadcast message
and deposits either in the individual task mailbox or in the group mailbox or in the
mailboxes of all tasks as in the code below. It serves as an interface between the data
link layer and the application and also serves the group.

```
/***************************************************************
                   Transport Layer Manager
 ***************************************************************/
TASK TL_Man() {

    data_unit_ptr_t data_pt;
    task_ptr_t task_pt;
    ushort tid, mbox_id3;
    ushort mbox_id1=task_tab[current_tid].mailbox_id;
    ushort mbox_id2=task_tab[System_Tab.DL_Out_id].mailbox_id;

    while(TRUE) {
      data_pt=recv_mailbox_wait(mbox_id1);
      if (data_pt->TL_header.form==UNICAST){
        tid=(data_pt->TL_header.receiver_id) & 0x00FF;
        mbox_id3=task_tab[tid].mailbox_id;
        send_mailbox_notwait(mbox_id3,data_pt);
      }
      else if (data_pt->TL_header.form==MULTICAST) {
        group_id=data_pt->TL_header.receiver_id;
        mbox_id2=group_tab[group_id].mailbox_id;
        send_mailbox_notwait(mbox_id2,data_pt);
      }
      else if (data_pt->TL_header.form==BROADCAST) {
        for(i=0;i<System_Tab.N_TASK;i++)
```

```
            if (task_tab[i].state=ALLOCATED) {
               data_pt2=get_data_unit(System_Tab.Net_Pool);
               memcpy(data_pt2, data_pt1, sizeof(data_unit_t));
               send_mailbox_notwait(task_tab[i].mailbox_id,data_pt);
            }
        }
}
```

Reception of a message is performed by assembling the received blocks into the message. We have a single blocking receive routine for unicast and multicast receptions with the type specified as a parameter passed to this routine as shown in the code below.

```
/*******************************************************************
                 receive  message by blocking
 ******************************************************************/
int recv_wait(char* msg_pt, ushort len, int type) {
    task_ptr_t task_pt;
    data_unit_ptr_t data_pt;
    ushort node_id, mbox_id, group_id;

    if (type == MULTICAST) {  // receive multicast
       group_id=task_tab[current_tid].group_id;
       mbox_id=group_tab[group_id].mailbox_id;
    }
    else mbox_id=task_tab[current_tid].mailbox_id;

    while(len>0) {
       data_pt=recv_mailbox_wait(mbox_id);
       memcpy(msg_pt, data_pt->data.data, N_DATA_UNIT);
       len=len-N_DATA_UNIT;
       msg_pt=msg_pt+len;
       put_data_unit(Sys_Tab.Net_Pool,data_pt);
    }
    return(DONE);
}
```

6.4.3 Data Link Layer Interface Tasks

We have two tasks to implement data link layer interface; *DL_Out* and *DL_In* in line with what is described in Sect. 6.2.2. Additionally, we will assume the basic error handling and flow control functions are handled by the data link layer tasks and the real-time network protocol handles only network communications at physical layer. The simple protocol we will implement is called the *Stop-and-Wait* protocol in which the sender waits for acknowledgement for every frame it sends. The FSM representation of this protocol is shown in Fig. 6.12.

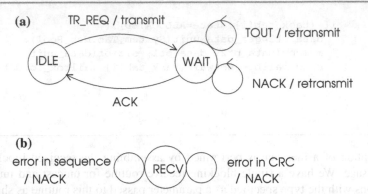

Fig. 6.12 Data link layer FSMs. **a** The sender. **b** The receiver

Table 6.1 FSM table for the data link layer sender

	Inputs |	TR_REQ	ACK	NACK	TOUT
States	IDLE	*act*_00	NA	NA	NA
	WAIT	NA	*act*_11	*act*_12	*act*_13

The sender FSM has two states; IDLE, when there is no data to transmit and WAIT when a frame is sent and an ACK message waits. If there is a NACK message meaning that the frame sent is received with error, or a time-out, the frame is retransmitted and a reply is waited. Sending is repeated for a predetermined number of times before it is aborted. The FSM table for the sender is depicted in Table 6.1. Note that some actions are not applicable (NA). For simplicity, we assume the transport layer does not request to send the next frame by the TR_REQ command until the current frame is delivered correctly. The routines *CRC_check* and *set_timer* which can be implemented simply are not shown.

This FSM is realized by the *DL_Out* task code of which is shown below. All data link tasks are in file "datalink.c".

```
/****************************************************************
                    Data Link Output Task
****************************************************************/
// datalink.c
#define IDLE          0  // states
#define WAIT          1
#define DATA_MSG      0 // message types
#define ACK           1
#define NACK          2
#define TOUT          3
#define N_TRIES_MAX   5

  fsm_table_t sender_FSM[2][4];
  data_unit_ptr_t data_pt;
```

```
  ushort dev_id, mbox_id, seq_no=0, crc_code, n_tries=0;

void act_00(){
  data_pt->MAC_header.type=DATA_MSG;
  data_pt->MAC_header.seq_num=seq_no;
  CRC_generate(data_pt1, &crc_code);
  MAC_trailer=crc_code;
  write_dev(dev_id,data_pt,N_DATA_LEN);}

void act_11(){
  daat_ptr_t data_pt2=get_data_unit(System_Tab.Net_Pool);
  data_pt2->type=ACK;
  seq_no=(seq_no+1) MOD 2;
  write_dev(System_Tab.Net_Dev,data_pt,N_DATA_LEN);}

void act_12(){
 if (++n_tries< N_TRIES_MAX) {
  write_dev(System_Tab.Net_Dev,data_pt,N_DATA_LEN);
  set_timer();
 else // error log
 } }

TASK DL_Out() {

    sender_FSM[0][0]=act_00;
    sender_FSM[1][1]=act_11;
    sender_FSM[1][2]=act_12;
    sender_FSM[1][3]=act_12;
    dev_id=Sys_Tab.Net_Dev;
    mbox_id=&(task_tab[current_pid])->mailbox_id);
    current_state=IDLE;

    while(TRUE)
    { data_pt=recv_mailbox_wait(mbox_id);
      (*sender_FSM[current_state][data_pt->MAC_header.type])();
    }
}
```

The reception of a message from the network is performed by the *DL_In* task which has only RECV state. It always waits for frames from the physical layer and upon reception, checks for errors and sends an acknowledgement or a negative acknowledgement accordingly.

```
/*****************************************************************
                    Data Link Input Task
*****************************************************************/

TASK DL_In() {

    data_unit_ptr_t data_pt1, data_pt2;
    ushort dev_id, mbox_id, crc_code, seq_no=0;
```

```
      dev_id=System_Tab.Net_Dev;
      mbox_id=task_tab[current_pid].mailbox_id;
      while(TRUE)
      {  data_pt=read_device(dev_id,System_Tab.DUNIT_LEN);
         CRC_generate(data_pt, &crc_code);
         if ((data_pt->MAC_header.seq_num != seq_no) ||
              (crc_code != data_pt->MAC_trailer))
            data_pt-> MAC_header.type= NACK;
         else {
            data_pt1->MAC_header.type= ACK;
            mbox_id=task_tab[System_Tab.TL_Man_id].mailbox_id;
            send_mbox_notwait(mbox_id,data_pt);
            type=data_pt1->TL_header.form;
            if (type==MULTICAST||type==BROADCAST) { // forward
              data_pt2=get_data_unit(System_Tab.Net_Pool);
              memcpy(data_pt2, data_pt1, System_Tab.N_DATA_UNIT);
              mbox_id=task_tab[System_Tab.DL_Out_id].mailbox_id;
              send_mbox_notwait(mbox_id,data_pt2)
          }
          }
        write_device(dev_id,data_pt,N_DATA_UNIT);
        seqno=(seqno+1) MOD 2;
      }
}
```

6.4.4 Group Management

We will start first by defining the *group control block* (*gcb*) for group management defined in the file "group.h" below. The group management basic routines are for creating a group, joining a group, and leaving a group. The sending of a multicast message to group members is achieved by the general *send* routine in the transport layer interface. Reception of a message is performed by each task from the group mailbox in return, again by the general *receive* function.

```
/***************************************************************
                  group data structure
 ***************************************************************/
/* group.h */
 #define ERR_GR_NONE    -2
 #define N_MEMBERS      30

typedef struct group  { ushort id;
                         ushort state;
                         ushort mailbox_id;
                         ushort n_members;
                         ushort local_members[N_MEMBERS];
                        }group_t;
typedef group_t* group_ptr_t;
```

A group structure is first allocated by the *allocate_group* system call as follows:

```
/************************************************************
                 allocate a group
************************************************************/

int allocate_group() {
    int i;

    for(i=0; i< System_Tab.N_GROUP; i++)
        if (group_tab[i].state != ALLOCATED) {
            group_tab[i].state= ALLOCATED;
            group_tab[i].id= i;
            group_tab[i].mailbox_id=make_mailbox();
            group_tab[i].n_members=1;
            group_tab[i].local_members[0]=current_tid;
            return(i);
        }
    }
    return(ERR_NOT_ALLOC);
}
```

A task can join an already allocated group as follows:

```
/************************************************************
                   join a group
************************************************************/

int join_group(ushort group_id) {
    group_ptr_t group_pt=&group_tab[group_id];

    if (  group_pt->state == ALLOCATED) {
        group_pt->n_members++;
        group_pt->local_members[n_members]=current_tid;
        return(DONE);
    }
    else return(ERR_GR_NONE);
}
```

Leaving a group by a task is performed by the *leave_group* system call.

```
/************************************************************
                   leave a group
************************************************************/

int leave_group(ushort group_id) {
    group_ptr_t group_pt=&group_tab[group_id];

    if (group_pt->state == ALLOCATED) {
        for(i=0;i<n_members;i++)
          if(group_pt->local_members[i]=current_pid)
```

```
            group_pt->local_members[i]=NOT_ALLOC;
            group_tab[i].n_members--;
            return(DONE);
        }
    else return(ERR_GR_NONE);
}
```

6.4.5   Clock Synchronization Algorithm

We will implement the NTP for distributed clock synchronization middleware of
DRTK. This task runs in a loop by delaying itself for 10 s first. It then generates a
message to ask time from the NTP server. It records the time the outgoing message
is sent and the time when the message from the server is received. Time offset can be
calculated based on these values and the timestamps (t_2 and t_3) stored in the message
by the server. We have a single code for the server and the clients. For simplicity,
we implement the task for one server and one time reading only without averaging
the values taken from different servers for a number of times. Since our simulation
environment is UNIX, we use the UNIX function $gettimeofday$ to read the current
time. The times t_2 and t_3 in the original algorithm are assumed to be equal. Since we
do not want to set the clock time of UNIX, the set_my_clock routine is left for a real
application.

```
/*************************************************************
                        NTP Task
*************************************************************/
/* clocksynch.c */
#include <time.h>
#include <sys/time.h>

#define CLOCK_READ     12

TASK NTP() {

    data_unit_ptr_t data_pt;
    ushort dest_id;
    struct timeval t1, t2, t4;
    long int t_offset;

  while(TRUE) {
   if(System_Tab.this_node=System_Tab.NTP_id) {
    data_pt=recv_mbox_wait(task_tab[current_tid].mailbox_id);
    dest_id=data_pt.TL_header.sender_id;
    data_pt->TL_header.sender_id=current_tid;
    data_pt->TL_header.receiver_id=dest_id;
    gettimeofday(&t2,NULL);
    data_pt->data.timestamp=t2;
    send_mbox_notwait(Sys_Tab.DL_Out_mbox,data_pt);
   }
```

```
    else {
      delay_task(current_tid, 10000);
      data_pt=get_data_unit(Sys_Tab.Net_Pool);
      data_pt->TL_header.type=CLOCK_READ;
      gettimeofday(&t1,NULL);
      data_pt->data.timestamp=t1;
      send_mbox_notwait(Sys_Tab.DL_Out_mbox,data_pt);
      data_pt=recv_mbox_wait(task_tab[current_tid].mailbox_id);
      gettimeofday(&t4,NULL);
      t2=data_pt->data.timestamp;
      t_offset=((t2.tv_sec*1e6 + t2.tv_usec)-(t1.tv_sec*1e6
            + t1.tv_usec)+(t4.tv_sec*1e6 + t4.tv_usec)-
          (t2.tv_sec*1e6 + t2.tv_usec))/2;
      set_my_clock(t2+t_offset);
    }
  }
}
```

6.4.6 Leader Election in a Ring

We will implement the leader election algorithm described in Sect. 6.3.3.1 in a ring structure. The FSM table for the depicted algorithm is shown in Table 6.2.

Note that we do not have any action at LDR state since this is a terminal state reached only by the new leader. The code below can be used to implement this FSM. The leader election task (*Leader_Elect*) at each node is awaken by a received message from its mailbox and acts according to the actions. We do not explicitly specify the mechanism for time-out to detect that the current leader as down since it depends on the application. Note that *act_11* has the same code as *act_01* since receiving an election message from IDLE state or ELECT state causes the same execution sequence. We assume the data area of the data unit is a union of other structures and the new field *winner_id*, which holds the highest node identifier that the message has traversed in the network.

Table 6.2 FSM table for the data link layer sender

		Inputs		
		TOUT	ELECTION	LEADER
States	IDLE	*act_*00	*act_*01	NA
	ELECT	NA	*act_*11	*act_*12
	LDR	NA	NA	NA

```
/**************************************************************
                    Leader Election Task
**************************************************************/
/* datalink.c */

 #define IDLE        0 // states
 #define ELECT       1
 #define LDR         2
 #define TOUT        0 // inputs
 #define ELECTION    1
 #define LEADER      2

fsm_table_t leader_FSM[3][2];

ushort recvd_id, my_leader, winner_id;
data_unit_ptr_t data_pt;

void act_00(){
    current_state=ELECT;
    data_pt=get_data_unit(System_Tab.Net_Pool);
    data_pt->TL_header.type=ELECTION;
    data_pt->TL_header.data.winner_id=System_Tab.this_node;
    send_net(data_pt, 0, BROADCAST);
}

void act_01(){
    current_state=ELECT;
    recvd_id=data_pt->data.winner_id;
    if (recvd_id < System_Tab.this_node)
        data_pt->data.winner_id=System_Tab.this_node;
    else if (recvd_id == System_Tab.this_node) {
        current_state=LDR;
        data_pt->data.TL_header.type=LEADER;
    }
    send_net(data_pt, 0, BROADCAST);
}

void act_12(){
    current_state=IDLE;
    my_leader=data_pt->data.winner_id
}

TASK Leader_Elect() {
    current_state=IDLE;
    sender_FSM[0][0]=act_00;
    sender_FSM[0][1]=act_01;
    sender_FSM[1][1]=act_01;
    sender_FSM[1][2]=act_12;

    while(TRUE)
    { data_pt=recv_mbox_wait();
      (*leader_FSM[current_state][data_pt->TL_header.type])();
    }
}
```

6.5 Review Questions

1. What are the key functions of a DRTOS?
2. Describe the main functions of the transport layer of the DRTOS.
3. What is the aim of the data link layer?
4. What is middleware and what are the main functions realized by the middleware of a DRTS? Give examples of common middleware modules in a DRTS.
5. What is the main application area of real-time task groups?
6. Why is clock synchronization needed in a DRTS or a non-real-time distributed system?
7. What is the difference between the Cristian's algorithm and the Berkeley algorithm?
8. Compare RBS and TPSN methods of clock synchronization used in WSNs.
9. What are the main usages of leaders in a WSN?
10. What are the common procedures of leader election in a WSN?

6.6 Chapter Notes

We reviewed the basic functions of a DRTOS and the real-time middleware in this chapter. A fundamental task to be performed by the DRTOS is the interface to the network, which is commonly achieved at transport and data link layers. The transport layer routines basically divide the user message into a number of data link layer protocol units at the sender side and assembling of the message is performed at the receiver side. Some or all of the functions at data link layer may be realized by the real-time protocol. We assumed the protocol provides only the physical layer functions and provided data link tasks to handle network procedures at this layer as DRTOS tasks. These tasks communicate and synchronize like any other task of the operating system and have higher priority than other tasks as delaying of them may result in missing deadlines of real-time application tasks. We showed the actual code to be implemented on top of the sample DRTK to realize data link layer tasks.

The real-time middleware consists of software modules, which are not typically part of the DRTOS but are needed by many diverse applications and hence, there is a need for these modules. We reviewed three such middleware modules; task group management, clock synchronization, and leader election. Out of these, clock synchronization is needed in many systems, whether real-time or not. Task groups provide a simple and elegant way to achieve fault tolerance in a DRTS, by running all copies of a task in parallel in a group and activating a copy when the primary task fails. We will see how message ordering can be achieved in Chap when fault tolerance in a DRTS is reviewed. Leader election is needed when the task set is partitioned into a number of subsets. A leader elected in a subset eases the functions to be performed in the subset, as in the case of task groups.

6.7 Programming Projects

1. Design and implement in C code the algorithm for a *send* routine at transport layer that waits for a reply from the sender. Implement the corresponding *receive* routine of the transport layer to be interfaced to DRTK.
2. Write the code for the Berkeley clock synchronization algorithm of the master and the worker nodes in C to be interfaced to DRTK.
3. Work out the FSM of the Bully Algorithm of Sect. 6.3.3.1 and write its code in C to be interfaced to DRTK.
4. Design and implement the FSM-based code of the TPSN protocol task in C to be interfaced to DRTK. Write both the level discovery phase and synchronization phase for the sink node and an ordinary node of the network.

References

1. Cheriton DR, Zwaenepoel W (1985) Distributed process groups in the V kernel. ACM Trans Comput Syst 3(2):77–107
2. Cristian F (1989) Probabilistic clock synchronization. Distrib Comput 3(3):146–158 (Springer)
3. Erciyes K (1989) Design and implementation of a real-time multi-tasking kernel for a distributed operating system. PhD thesis, Computer Eng. Dept, Ege University
4. Elson, J, Estrin D (2002) Fine-grained network time synchronization using reference broadcast. In: The fifth symposium on operating systems design and implementation (OSDI), pp 147–163
5. Fahmy S, Younis O (2004) Distributed clustering in ad-hoc sensor networks: a hybrid, energy-efficient approach. In: Proceedings of the IEEE conference on computer communications (INFOCOM), Hong Kong
6. Ganeriwal S, Kumar R, Srivastava M (2003) Timing-Sync protocol for sensor networks. In: The first ACM conference on embedded networked sensor systems (SenSys), pp 138–149
7. Heinzelman WR, Chandrakasan A, Balakrishnan H (2000) Energy-efficient communication protocol for wireless microsensor networks. In: Proceedings of the 33rd annual Hawaii international conference on system sciences, vol 2, p 10
8. Heinzelman WR, Chandrakasan A, Balakrishnan H (2002) An application-specific protocol architecture for wireless microsensor networks. IEEE Trans Commun 1(4):660–670
9. Mills DL (1992) Network time protocol (version 3): specification, implementation, and analysis. RFC 1305
10. Mills DL (2010) Computer network time synchronization: the network time protocol. Taylor & Francis, p 12. ISBN 978-0-8493-5805-0
11. Gropp W, Lusk E, Skjellum A (1999) Using MPI: portable parallel programming with the message passing interface, 2nd edn. MIT Press
12. Sivrikaya F, Yener B (2004) Time synchronization in sensor networks: a survey. IEEE Netw 18(4):45–50
13. Simple Network Time Protocol (SNTP) Version 4 for IPv4, IPv6 and OSI. RFC 4330
14. Zogg J-M (2002) GPS basics. Technical Report GPS-X-02007, UBlox, Mar 2002

Part III
Scheduling and Resource Sharing

Uniprocessor-Independent Task Scheduling

7.1 Introduction

Scheduling is the process of assigning tasks to a processor or a set of processors as in the case of a multiprocessor system or to a network of computing elements when a distributed real-time system is considered. A real-time task has a release time, an execution time, a deadline, and resource requirements. A periodic task is activated at regular intervals, an aperiodic task may be activated at any time, and a sporadic task has a minimum interval between any of its consecutive activations. Tasks may have precedence relation among them which means a task cannot start before its predecessors complete. Moreover, tasks may be sharing resources which affects scheduling decisions. Our main goal in scheduling real-time tasks is to have tasks meet their deadlines and provide a fair sharing of resources.

We will distinguish between scheduling independent tasks, dependent tasks with no resource sharing, dependent tasks with resource sharing, multiprocessor scheduling of independent and dependent tasks, and finally distributed scheduling of independent and dependent tasks. Each of these different applications may require somewhat diverse scheduling methods. Our approach is to classify these methods as uniprocessor-independent task scheduling; uniprocessor-dependent and resource-sharing task scheduling; and multiprocessor and distributed task scheduling. Using this classification, we have a chapter for each scheme starting with uniprocessor-independent task scheduling in this chapter.

We begin by first reviewing main concepts related to real-time scheduling. We then describe the main scheduling policies followed by the fundamental methods of scheduling in a uniprocessor system. Finally, we show how various scheduling policies can be adopted in the sample kernel DRTK.

© Springer Nature Switzerland AG 2019
K. Erciyes, *Distributed Real-Time Systems*, Computer Communications and Networks, https://doi.org/10.1007/978-3-030-22570-4_7

7.2   Background

Types of tasks and their attributes are fundamental in deciding what scheduling method to use. We can have the following types of real-time tasks in general:

- *Periodic Tasks*: A periodic task τ_i is activated at regular intervals called its *period* (T_i). Most of the hard real-time tasks are in this category. For example, monitoring temperature of a chemical plant can be performed by a task that is activated every few seconds.
- *Aperiodic Tasks*: These are the tasks that are activated at unpredictable times. Commonly, an aperiodic task is activated by an external interrupt to the real-time system such as a user pressing a button on the control panel. An alarm activating task that notices a parameter is out of range in a process control system is also an example of an aperiodic task.
- *Sporadic Tasks*: These tasks need to start at irregular times as the aperiodic tasks but a minimum time between each arrival of such a task is known beforehand.

Tasks of a real-time system may have the following attributes:

- *Arrival Time a_i*: This is the time when a task becomes ready for execution. Also called *release time* or *request time* (r_i).
- *Worst-Case Execution Time* (WCET) C_i: This is the estimate of the worst (longest) possible execution time of the task τ_i.
- *Start Time s_i*: The time at which task τ_i starts execution.
- *Finish Time f_i*: The time at which task τ_i finishes execution.
- *Response time R_i*: The time interval between the release time of a task and the time it finishes, $f_i - r_i$.
- *Task Period T_i*: Constant interval between successive activations of a periodic task.
- *Absolute Deadline d_i*: The absolute time task τ_i must finish execution.
- *Relative Deadline D_i*: The time interval between the arrival time and the time task T_i must finish execution.
- *Slack Time S_i*: This is the time between the relative deadline of the task and its worst execution time as $D_i - C_i$.
- *Laxity Time $L_i(t)$*: The remaining execution time of a task at time t.
- *Time Overflow*: This occurs when a task finishes after its deadline.

All these parameters are shown in Fig. 7.1. While attempting to schedule tasks with deadlines, our aim is to provide an assignment with no time overflows in which case the schedule is called *feasible*. A *job* of a task is commonly used in literature to denote the instant of a task that is activated. We will refer to the running instant of a task simply as *task* to avoid confusion with the operating system job concept.

Fig. 7.1 Task attributes

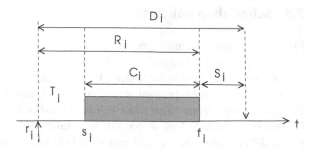

7.2.1 Schedulability Test

If the main task characteristics of tasks are known in advance, we can perform the following schedulability tests:

- *Necessity*: If this test is not successful, the task set is not schedulable. If the test succeeds, we have the necessary condition but cannot be sure that the task set is schedulable. Therefore this test is mainly used to label a task set as unschedulable.
- *Sufficiency*: In this case, passing this test means that the task set is schedulable. However, a negative outcome from this test does not mean the set is unschedulable.
- *Exactness*: We have both of the above tests having positive results.

7.2.2 Utilization

The utilization factor μ_i (or u_i) for a periodic task τ_i is defined as follows:

$$u_i = \frac{C_i}{T_i} \tag{7.1}$$

In other words, this parameter shows the percentage of time a task uses the processor during its period. The processor utilization factor U is the fraction of time spent in executing a task set $\mathcal{T} = \tau_1, \ldots, \tau_n$ which is

$$U = \frac{C_1}{T_1} + \cdots + \frac{C_n}{T_n} \tag{7.2}$$

It can be seen a better processor utilization factor can be obtained by increasing the computation time or decreasing the periods of tasks as long as the assignment remains feasible.

7.3 Scheduling Policies

First and foremost, we must provide a meeting of deadlines of hard real-time tasks in any scheduling method and if we are not able to do so, we should not allow a task with an unmet deadline to the system. If tasks share resources, the operating system should provide efficient protection mechanisms of these resources as we reviewed in Chap. 4. Another point to consider is whether tasks are dependent or independent. Moreover, we need to look at ways of scheduling a combination of hard, firm, and soft real-time tasks which reflect real-world phenomenon.

7.3.1 Preemptive or Non-preemptive Scheduling

In a non-preemptive scheduling policy, a task dispatched is run to completion without being interrupted. Two common policies adopted in the batch processing form of this method is as follows:

- *First Come First Served* (FCFS) Scheduling: Tasks are placed in a dispatch queue with respect to their arrival times and assigned to the processor from the front of this queue in turn.
- *Shortest Job Next* (SJN) Scheduling: A shorter task will be scheduled prior to a longer executing task in this policy. We need to know the execution times of tasks before running them in this policy and this may not be feasible for a range of real-time tasks. Furthermore, a task with a very long execution time may be starved of the processor for a long time by many short executing tasks.

On the other hand, a preemptive scheduling policy allows the interruption of a running task to let a higher priority task to run. Let us consider a sample task set of three tasks shown in Table 7.1 with P_i being the priority of a task τ_i.

If we assign static higher priorities to tasks with shorter periods and employ non-preemptive scheduling policy, we will have the scheduling as shown in Fig. 7.2. It can be seen that task τ_1 cannot complete its execution before its deadline at the start of its second period at time 16.

However, had we implemented a preemptive policy, we could have the assignment in Fig. 7.3 where each task meets its deadline. Task τ_3 is preempted at time 8 to schedule task τ_1, and then at time 12 to schedule τ_2 which is preempted at time 16 by τ_1 so that all meet their deadlines. Note that we can repeat this major cycle of 24

Table 7.1 An example task set

τ_i	C_i	T_i	P_i
1	2	8	1
2	5	12	2
3	8	24	3

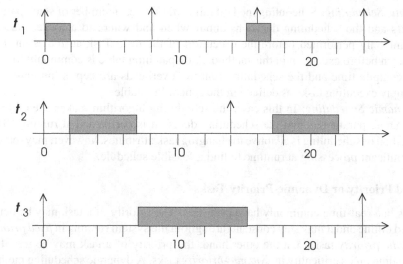

Fig. 7.2 Non-preemptive scheduling of the three tasks of Table 7.1

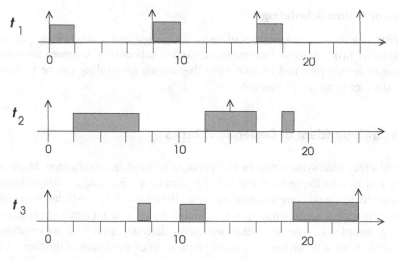

Fig. 7.3 Preemptive scheduling of the three tasks of Table 7.1

time units as all tasks are periodic. Different ways of assigning priorities give rise to different scheduling policies as we will describe in the next sections.

7.3.2 Static or Dynamic Scheduling

We have two types of scheduling methods considering the time of scheduling decisions taken as follows:

- *Static Scheduling*: Scheduling method (algorithm) uses a number of static parameters and the scheduling decisions about when and where to activate a task is commonly performed before the execution of the task. Task attributes must be known before execution in this method. A dispatching table is commonly formed at compile time and the scheduling runtime overheads are kept at minimum by simply executing tasks as defined in the scheduling table.
- *Dynamic Scheduling*: In this case, the scheduling algorithm makes use of some dynamic parameters and the scheduling decision is performed at runtime. This method of scheduling is adaptive to changing task attributes, however, may require significant processing at runtime to find a feasible schedule.

Fixed Priority or Dynamic-Priority Tasks

Tasks in a real-time commonly have priorities. The priority of a task may be determined offline and it may stay constant throughout the system running in *fixed priority* or *static priority* tasks. On the other hand, the priority of a task may be modified depending on its criticality in *dynamic-priority* tasks. A dynamic scheduling method may have tasks with static deadlines as we will see.

Offline or Online Scheduling

In *offline scheduling*, assignments of tasks is decided and a schedule is generated for the entire task set before running tasks. Online scheduling decisions are made at runtime whenever new tasks arrive. Note that online scheduling can be performed using static or dynamic parameters.

7.3.3 Independent or Dependent Tasks

Tasks of a real-time system may be independent without any interaction. More commonly, real-time tasks need to communicate to transfer some data and to synchronize, for example, to signal completion of an event. When a task τ_i sends data to a task τ_j, we say τ_i precedes τ_j, shown as $\tau_i \prec \tau_j$, which is to say that τ_j cannot start execution before τ_i completes. Note that τ_i may be sending data at its first 10% of execution but we assume τ_i must finish before τ_j starts running since detection of the time of data sending is not practical. In general, all of the predecessors of a task must finish execution before it can begin. This precedence relation is depicted in a *task dependency graph* (or simply a *task graph*) where an arc pointing from τ_i to τ_j shows $\tau_i \prec \tau_j$ as displayed in Fig. 7.4. There are seven tasks, $T_1, \ldots, T_7$ and T_7 is the terminal task which has to wait, until all other tasks finish execution. The communication between two tasks may have nonzero cost when these two tasks are on two different nodes of a distributed real-time system. However, the communication costs are assumed to be negligible when tasks run on the same processor since sending of data would involve passing the address of data and possibly signaling this event to the receiver of the message.

Another point to note is that the execution times and precedence relationships of tasks should be known before we construct the task graph, hence, we need to

Fig. 7.4 A task graph with task execution times shown next to them

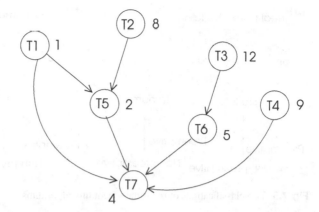

perform *static scheduling* for these tasks. Therefore, an important aim of any task scheduling algorithm is to obey the precedence relationships between tasks in this case. Moreover, when we aim to distribute tasks to the nodes of a distributed real-time system as we will see in Chap. 9, we need effective heuristic algorithms to partition the task graph.

The scheduling problem becomes more difficult when tasks share resources. Shared resources are protected by operating system constructs such as semaphores and locks, however, scheduling of tasks when they compete for a resource requires careful analysis as a resource required by a task may not be available during its execution. In such a case, execution of a task may be delayed causing miss of a deadline or even a deadlock may occur as we will review in the next chapter.

7.4 A Taxonomy of Real-Time Scheduling Algorithms

We can classify scheduling algorithms in a real-time system using various approaches. We will first assume that tasks are independent and our aim is to find a feasible schedule for tasks to meet their deadlines in a uniprocessor system. Two different classifications of real-time scheduling algorithms proposed in (a) [4] and (b) [1] are depicted in Fig. 7.5.

We see the main difference is whether static or dynamic scheduling or periodic or aperiodic tasks are considered initially for hard real-time tasks. We will see these are dependent, in fact, for example, many hard real-time tasks are periodic and have static priorities at the same time. In the next sections, we will review fundamental real-time scheduling algorithms with assumptions that there are no interactions between tasks and our aim is to obtain a feasible schedule on a single processor. We will assume the following most of the time when searching for a scheduling algorithm. Scheduling algorithms for dependent real-time tasks that share resources will be reviewed in the next chapter.

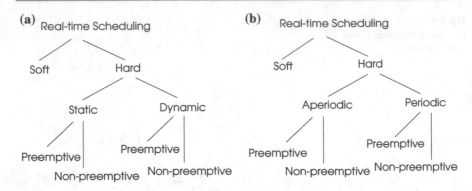

Fig. 7.5 Two classifications of real-time scheduling algorithms

- The task set $T = \{\tau_1, \ldots, \tau_n\}$ to be scheduled consists of periodic tasks only.
- Deadline D_i of a task τ_i is equal to its period T_i.
- Computation time C_i of a task τ_i is determined beforehand and remains constant.

7.5 Clock-Driven Scheduling

We will first consider a simple model in which tasks are periodic, independent, they do not share any resources and their priorities are fixed. A typical application with such characteristics consists of a set of sensors which are activated at regular intervals (periods), sense data and send this data to the input unit of the real-time computer upon which a real-time task is invoked to process the input data. This processing is to be completed before the next data is input from the sensor. Scheduling points are known in advance and therefore, a table that specifies when to invoke the scheduler can be specified. In this so-called *clock-driven* method, two basic approaches are commonly used; table-driven (or cyclic scheduling) and cyclic executive (or structured cyclic scheduling) algorithms.

7.5.1 Table-Driven Scheduling

Table-driven cyclic scheduling is a simple offline method that has been used in industry for decades for hard real-time periodic tasks. It is based on a pre-calculated table that contains the execution times of tasks. This type of scheduler, in fact, does not need an operating system support, a hardware timer that expires at task scheduling points is sufficient. Moreover, calculation of table entries may involve complex algorithms requiring significant times as this is performed offline. We assume there are n periodic tasks with known release times and worst-case execution times in this

Table 7.2 An example task set

τ_i	C_i	T_i
1	1	4
2	2	10
3	2	10
4	4	20

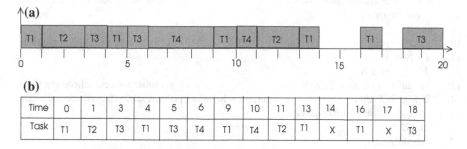

(a)

T1	T2	T3	T1	T3	T4	T1	T4	T2	T1		T1		T3

0   5   10   15   20

(b)

Time	0	1	3	4	5	6	9	10	11	13	14	16	17	18
Task	T1	T2	T3	T1	T3	T4	T1	T4	T2	T1	X	T1	X	T3

Fig. 7.6 Clock-driven scheduling, **a** execution of four tasks of Table 7.1, **b** table contents

algorithm to be able to form the task table offline. Let us consider a sample task set of four tasks $\tau_1, \ldots, \tau_4$ with computation times and periods shown in Table 7.2.

A possible schedule of these tasks is depicted in Fig. 7.6. Note that after 20 time units, the schedule repeats, this period is the lowest common multiple of all task periods. We assumed deadlines of tasks equal their periods, that is, each task must complete its execution in its current period before its next period starts, and preemption is possible with insignificant context switching times. The points in the table marked with X are idle times, which can be used for activation of any aperiodic task that is waiting in the background.

Greatest common divisor (GCD) of task periods is termed the *minor cycle* and lowest common multiple of periods LCMs of task periods are the *major cycle*.

Remark 7.1 We need only consider switching of tasks at GCD points.

Remark 7.2 Once we obtain a feasible schedule for the major cycle, this assignment will repeat in cycles and hence, the name cyclic scheduler.

The scheduling algorithm of the cyclic scheduler is shown in Algorithm 7.1 [3]. It waits for interrupts only at context switching points and activates the task that is specified in the table in that scheduling point. Note that we allow execution of aperiodic tasks in this version of the cyclic scheduler. If there are no anticipated aperiodic tasks in the system, lines 7–9 and 14–16 of the algorithm may be omitted and table should be modified so as not to have any scheduler activation at idle points of time. This would mean deleting activations at times of 14 and 17 from the table.

Algorithm 7.1 *Table_Scheduler*

1: **Input:** set $T = \{\tau_1, \ldots, \tau_n\}$ of n periodic tasks
2: aperiodic task queue AT_queue
3: $i \leftarrow 0, k \leftarrow 0$
4: set timer to expire at t_k
5: **while** *true* **do**
6:     **wait** for timer interrupt
7:     **if** τ_{curr} is aperiodic **then**   ▷ if current task is aperiodic, preempt it
8:         preempt τ_{curr}
9:     **end if**
10:    $\tau_{curr} \leftarrow \tau(t_k)$ ▷ select task specified in the table
11:    $i \leftarrow i + 1$
12:    $k \leftarrow k \mod N$
13:    set timer to expire at $\lceil i/n \rceil H + t_k$ ▷ set timer for next scheduling point
14:    **if** τ_{curr} is X **then**   ▷ if current scheduling point is idle, select an aperiodic task
15:       $\tau_{curr} \leftarrow$ front of AT_queue
16:    **end if**
17:    schedule τ_{curr}
18: **end while**

7.5.2 Cyclic Executive

The table-driven scheduling works fine but as the number of tasks grow, the size of the table grows, which may not be convenient for embedded systems with limited memory. We can improve the operation of the table-driven cyclic scheduler by incorporating some structure in this method. In this modified approach to cyclic scheduling, time is divided into frames with constant sizes and scheduling decisions are made only at frame boundaries, hence, there are no preemptions of tasks inside a frame.

The size of the frame f should be selected with care. A frame should be large enough to have every task contained in it so that preemption within a frame is not needed. Hence, we have $f \geq \max(C_i)$, $i = 1, \ldots, n$ as the first constraint. On the other hand, f should divide the size of the hyper period $H = LCM(T_1, \ldots, T_n)$. Therefore, f divides T_i for at least one task τ_i,

$$\left\lceil \frac{T_i}{f} \right\rceil - \frac{T_i}{f} = 0$$

as the second constraint. Let $F = H/f$, the interval H is called a *major cycle* and the interval f a *minor cycle*. In order to provide meeting of deadlines of tasks, it can be shown that the following inequality must hold [3]:

$$2f - GCD(T_i, f) \leq D_i \tag{7.3}$$

which is the third constraint. Let us determine f for the tasks in Table 7.2. The first constraint is satisfied when $f \geq 4$. Hyper period H is 20 units, hence f can be 4, 5,

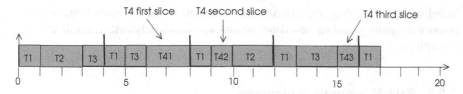

Fig. 7.7 Execution of four tasks of Table 7.1 by the cyclic executive

10, or 20 to divide H. The third constraint is not satisfied for all tasks in which case we can use a method called *task splitting*. We can see the longest executing tasks τ_4 which can be split into three subtasks τ_{4_1}, τ_{4_2}, and τ_{4_3} and we can now sketch a possible schedule as shown in Fig. 7.7.

The algorithm that implements the structured cyclic scheduling method is called a *cyclic executive*, which is initiated at timer interrupts at frame boundaries and executes the block as a whole until the next timer interrupt. The structure of this algorithm is depicted in Algorithm 7.2 again allowing the unused time to be used by aperiodic tasks [2].

Algorithm 7.2 *Cyclic_Executive*

1: **Input**: stored schedule $S = \{S_1, \ldots, S_k\}$, F
2: $i \leftarrow 0, t \leftarrow 0$
3: set *timer* to expire at t
4: **while** *true* **do**
5:    **wait** for *timer* interrupt
6:    *current_block* $\leftarrow S_i$
7:    $t \leftarrow t + 1$
8:    $i \leftarrow t \bmod F$
9:    **execute** all tasks in *current_block*
10:    **sleep** until next timer interrupt
11: **end while**

Cyclic scheduling in both forms, in basic or structured methods, is simple and efficient, however, even a minor change of task attributes may require a significant change of schedule that has to be computed from scratch. However, it continues to be used in many contemporary real-time applications due to its simplicity and low runtime overheads.

7.6 Priority-Based Scheduling

Online scheduling is based on making scheduling decisions during execution of tasks, it is therefore a dynamic approach to the scheduling problem. This method may involve static priority or dynamic-priority tasks. We will review four main priority-based online scheduling algorithms; rate monotonic, deadline monotonic as static priority algorithms, and earliest deadline first and least laxity first dynamic-priority algorithms.

7.6.1 Rate Monotonic Scheduling

The rate monotonic (RM) scheduling is a dynamic preemptive algorithm for static priority, periodic, and hard real-time tasks. We assume independent periodic tasks with deadlines equalling their periods. The main idea of this algorithm is to assign static priorities to tasks based on their periods, a shorter period meaning a higher priority. The rate of a task is the inverse of its period and hence the name of this algorithm. A higher priority task preempts a running task and context switching times are considered negligible.

Formally, a task τ_i has a higher priority than a task τ_j if $T_i < T_j$ assuming all tasks have distinct periods. When periods of two or more tasks are equal, task identifiers may be used to break symmetries. Let us consider the task set $\mathcal{T}$ in Table 7.3 with three tasks τ_1, τ_2, τ_3 with shown computation times and periods (deadlines). The priorities assigned to these tasks are shown in the last column of the table.

The utilization relationship to the number of tasks is given by the following theorem:

Theorem 7.1 (Liu [3]) *A set of n periodic hard real-time tasks will meet their deadlines independent of their start times if,*

$$\sum_i^n \frac{C_i}{T_i} \le n(2^{1/n} - 1) \tag{7.4}$$

Utilization factor U converges to $\ln 2 = 0.69$ as $n > 10$. We can therefore test whether a task set can be scheduled using this inequality. Note that this test is sufficient but not a necessary condition. In other words, there may be a task set that has a total utilization larger than 0.69 but tasks in this set may still meet their deadlines under RM scheduling.

Let us consider the task set $\tau_1(2, 8)$ and $\tau_2(8, 20)$ in $\tau_i(C_i, T_i)$ form for RM scheduling. We can see that these two tasks can be scheduled since $(2/8) + (8/20) = 0.65 < 2(2^0.5 - 1) = 0.82$. Hence, this task set can be admitted to the system. The scheduling of these tasks using RM scheduling is depicted in Fig. 7.8. Since LCM of their periods is 40, the schedule will repeat itself every 40 time units.

Table 7.3 A sample task set

τ_i	C_i	T_i	P_i
1	3	10	1
2	6	24	3
3	7	12	2

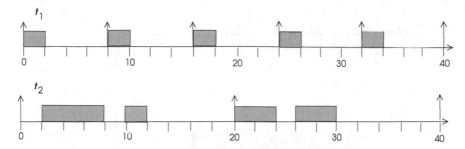

Fig. 7.8 Feasible RM scheduling of two independent periodic tasks that pass the admission test

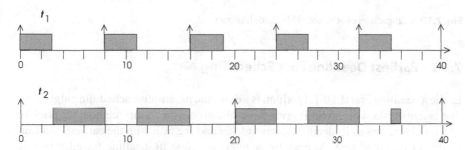

Fig. 7.9 Feasible RM scheduling of two independent periodic tasks that do not pass the test

Let us now increase the computation time of tasks to have $\tau_1(3, 8)$ and $\tau_2(10, 20)$. Applying the schedulability test yields $(3/8) + (10/20) = 0.875 > 0.82$, which is to say the RM algorithm does not guarantee to have a feasible schedule for this task set. However, we can still schedule these tasks to have their deadlines met as shown in Fig. 7.9.

The advantages of this method are that it is simple to implement and it is optimal among all static priority scheduling algorithms. However, it may not use the processing power efficiently as shown by the example above.

Deadline Monotonic Algorithm

There may be periodic real-time tasks that have deadlines different than their periods, that is, $\exists \tau_i : D_i < T_i$. In such cases, deadline monotonic (DM) algorithm which assigns priorities to tasks based on their deadlines only, can be used. The DM algorithm is a variation of the RM algorithm and may produce feasible schedules in cases where RM algorithm fails.

An example with two tasks as $\tau_i(C_i, D_i, T_i)$ is specified as follows: $\tau_1(2, 5, 5)$ and $\tau_2(3, 4, 10)$. Running the RM scheduling algorithm for these two tasks results in the schedule of Fig. 7.10a where τ_2 misses its deadline since τ_1 having a shorter period is scheduled first. These two tasks are scheduled using DM scheduling to give the schedule in Fig. 7.10b with each task running to completion within its deadline.

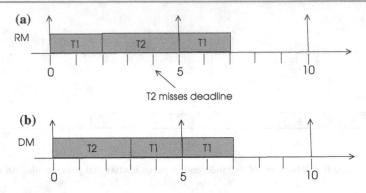

Fig. 7.10 Comparison of RM and DM scheduling methods

7.6.2 Earliest Deadline First Scheduling

Earliest deadline first (EDF) algorithm is a dynamic preemptive scheduling algorithm
that assigns tasks with dynamic priorities to the processor. In this method, the priority
of a task increases as its deadline comes closer and the task with the earliest deadline
is always executed first. At each timer interrupt, time to deadline for each task is
calculated and the task with the smallest value is dispatched. This frequent calculation
may produce considerable overhead when task set is large.

Remark 7.3 EDF will produce a feasible schedule if the utilization factor $U \leq 1$.

   This means EDF is an effective scheduling policy that can use 100% CPU power.
EDF may be implemented with both periodic and aperiodic tasks.

7.6.2.1 Aperiodic EDF

Let us consider the aperiodic case first with the task set given in Table 7.4. All three
tasks are aperiodic (one-shot) and have different absolute deadlines.

   The EDF schedule of this task set is shown in Fig. 7.11. We can see that task τ_1 is
preempted by task τ_2 at time 4 since the absolute deadline of τ_2 (14) is the earliest
deadline at that point in time. Later, when task τ_3 is released at time 8, there is no
preemption since τ_2 still has the nearest deadline. When τ_2 finishes at $t = 10$, τ_1

Table 7.4 A sample task set

τ_i	Arrival time a_i	C_i	Absolute deadline d_i
1	0	8	18
2	4	6	14
3	8	10	26

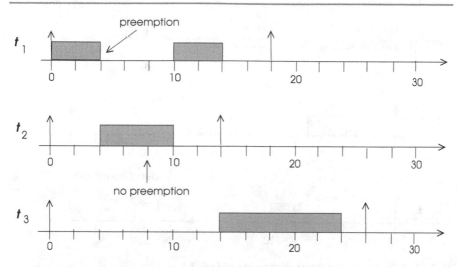

Fig. 7.11 EDF scheduling of three independent aperiodic tasks

Table 7.5 Example task set

τ_i	C_i	T_i
1	3	10
2	4	13
3	8	21

has the current earliest deadline and hence, it is scheduled. Task τ_3 has the farthest deadline and is scheduled after τ_2 to finish at time 24.

7.6.2.2   Periodic EDF

We now assume tasks are periodic, there is no precedence relation among them, there is no mutual exclusion requirement between any pair of tasks, the deadline of a task is equal to its period and context switching times are negligible as in the RM scheduling algorithm. The worst-case execution time for each task is constant and is known beforehand, again as in the RM algorithm.

Whenever a task becomes ready for execution, for example when its period starts, task priorities may be modified to assign the highest priority to the task that is closest in time to its deadline. The runtime scheduler than selects the task with the highest priority for execution. It can be shown that this method may achieve processor utilization up to 100% at the expense of adjusting priorities at runtime. Let us consider the task set of Table 7.5.

We will first attempt to schedule this set using the RM algorithm. The utilization factor is $(3/10) + (4/13) + (8/21) \approx 0.989$ which is more than $U = 3(2^{(1/3)} - 1) = 0.78$. However, we have seen that RM test provides sufficiency, in other words,

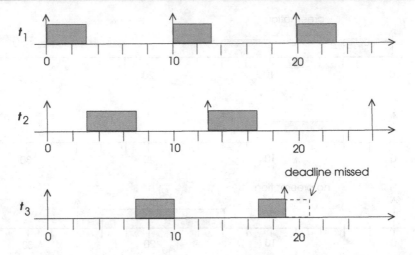

Fig. 7.12 RM scheduling of three periodic tasks of Table 7.5

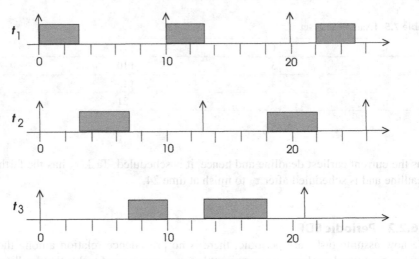

Fig. 7.13 EDF scheduling of three periodic tasks of Table 7.5

failing this test does not mean that the task set cannot be scheduled. The assignment obtained when we apply RM algorithm is shown in Fig. 7.12 and we can see that task τ_3 misses its deadline.

We now try to schedule this set using the EDF algorithm and the result is depicted in Fig. 7.13 which shows a feasible schedule. Note that task τ_3 does not need to wait for task τ_2 since it has a closer deadline and therefore has a higher dynamic priority.

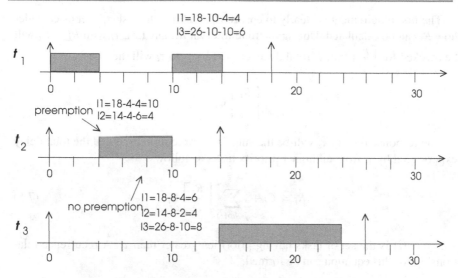

Fig. 7.14 LLF scheduling of the three aperiodic tasks of Table 7.4

7.6.3   Least Laxity First Algorithm

We defined *laxity* (*slack*) of a task as the difference between time to its deadline and its remaining execution time. Note that this parameter is dynamic since the remaining time of a task changes with time; the more it executes, the less its remaining time remains and hence less its current laxity. The *least laxity first* (LLF) (or the *least slack first*) algorithm evaluates laxity values for all tasks and schedules the task with the minimum of this value. Hence, LLF is a dynamic scheduling algorithm that assigns tasks with dynamic priorities as the EDF algorithm. Different than the EDF algorithm, the computation times of tasks affect the scheduling decisions. Let us consider the task set of Table 7.4 again. Using the LLF algorithm results in the schedule obtained in Fig. 7.14 with the shown laxity values obtained at critical instants. At time $t = 4$, τ_2 is activated and has a lower laxity value than τ_1 which results in preemption of τ_1. There is no preemption of τ_2 at $t = 8$ when τ_3 becomes available since τ_2 has the least laxity among all three tasks. The valid tasks at $t = 10$ are τ_1 and τ_2 and τ_1 has a lower laxity value and therefore is scheduled. We see this schedule is the same obtained by the EDF algorithm, however, these two algorithms may produce different schedules.

7.6.4   Response Time Analysis

The response time R_i of task τ_i is the time interval between the arrival of τ_i to the time it finishes execution. In a static priority-based system, $R_i = C_i + I_i$ where I_i is the delay caused by tasks that have higher priorities than τ_i.

The first requirement is clearly to ensure $R_i \leq D_i$ for the task τ_i. Let us consider how R_i can be calculated. During the time interval R_i, any task τ_j with $Pj > P_i$ will be invoked for $\left\lceil \frac{R_i}{T_j} \right\rceil$ times. Total delay caused by τ_j to τ_i will then be,

$$\left\lceil \frac{R_i}{T_j} \right\rceil C_j$$

The response time of τ_i will be the sum of its execution time and the total delay experienced by τ_i from all higher priority tasks as follows:

$$R_i = C_i + \sum_{j \in hp(i)} \left\lceil \frac{R_i}{T_j} \right\rceil C_j \tag{7.5}$$

where $hp(i)$ is the set of tasks having priorities greater than τ_i. A recurrence relationship for this equation can be formed,

$$R_i^{n+1} = C_i + \sum_{j \in hp(i)} \left\lceil \frac{R_i^n}{T_j} \right\rceil C_j \tag{7.6}$$

We can start the iteration with a value of R_i that is guaranteed to be less than or equal to its final value, for example $R_i^0 = C_i$, and stop the iteration when R_i^{n+1} converges to R_i^n or when $R_i^n > D_i$.

$$R_i = C_i + \sum_{j \in hp(i)} \left\lceil \frac{R_i}{T_j} \right\rceil C_j \tag{7.7}$$

Let us consider the task set of Table 7.6 with tasks of Table 7.1 ordered with respect to their priorities.

The response time for the highest priority task τ_1 is its computation time of 3 units since no other task will preempt it, and $R_1 \leq T_1$ meaning τ_1 will meet its deadline. Iterating for the response time of τ_3 yields,

$$R_3^1 = C_3 + \left\lceil \frac{R_3^0}{T_1} \right\rceil C_1 = 2 + \left\lceil \frac{2}{10} \right\rceil 3 = 5$$

Table 7.6 A sample task set

τ_i	C_i	T_i	P_i
1	3	10	1
3	2	12	2
2	6	24	3

Next iteration gives,

$$R_3^2 = 2 + \left\lceil \frac{5}{10} \right\rceil 3 = 5$$

Since there is no change in R_3 values, $R_3 = 5 \le T_3$ which shows τ_3 will execute within its deadline. We need to consider preemption by τ_1 and τ_3 while calculating response time for τ_2 as follows:

$$R_2^1 = C_2 + \left\lceil \frac{R_2^0}{T_3} \right\rceil C_3 + \left\lceil \frac{R_2^0}{T_1} \right\rceil C_1$$

$$6 + \left\lceil \frac{6}{12} \right\rceil 2 + \left\lceil \frac{6}{10} \right\rceil 3 = 11$$

Continuing with the iteration yields,

$$R_2^2 = 6 + \left\lceil \frac{11}{12} \right\rceil 2 + \left\lceil \frac{11}{10} \right\rceil 3 = 14$$

$$R_2^3 = 6 + \left\lceil \frac{14}{12} \right\rceil 2 + \left\lceil \frac{14}{10} \right\rceil 3 = 16$$

$$R_2^4 = 6 + \left\lceil \frac{16}{12} \right\rceil 2 + \left\lceil \frac{16}{10} \right\rceil 3 = 16$$

Since $R_2^4 = R_2^3$, we stop, $R_2 \le T_2$ and we can deduce all tasks will meet their deadlines which equal their periods. Note that this is a necessary and sufficient condition. Applying RM test,

$$U \le n(2^{1/n} - 1) = 3(2^{1/3} - 1) = 0.78$$

Since $U = (C_1/T_1) + (C_2/T_2) + (C_3/T_3) = 3/10 + 2/12 + 4/24 = 0.55$, we can say this task set is schedulable by RM scheduling.

7.7 Aperiodic Task Scheduling

A real-time system will inhibit a mixture of periodic and aperiodic tasks in general, necessitating the employment of a mechanism for aperiodic tasks. Aperiodic tasks are commonly associated with soft deadlines and sporadic tasks generally have hard deadlines. A simple approach to aperiodic task scheduling then involves using two task ready queues, higher priority queue for periodic hard real-time tasks and a lower priority queue for aperiodic soft real-time tasks as depicted in Fig. 7.15. Periodic tasks are scheduled using a priority-based algorithm such as RM, EDF or LLF. However, this simple scheme may suffer from possible long response times of aperiodic tasks when periodic task population is large.

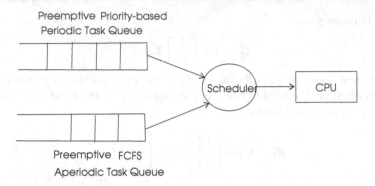

Fig. 7.15 Aperiodic task scheduling using two queues

7.7.1 Basic Approaches

Scheduling of aperiodic tasks should aim at their fast completion such that periodic task deadlines should also be met. In this context, the main goal of any aperiodic task scheduling algorithm is to minimize the average response times of aperiodic tasks without disturbing the guaranteed deadlines of periodic tasks. We can have two basic approaches when scheduling aperiodic tasks in a real-time system with periodic tasks.

- Aperiodic tasks may run in time intervals where the processor is idle. This is a simple method that works correctly, however, aperiodic tasks may suffer long delays causing significant average response times. Figure 7.16 depicts RM scheduling of tasks $\tau_1(0, 2, 5)$ and $\tau_2(0, 5, 15)$ with an aperiodic task $\tau_3(0, 1)$ the response time of τ_3 is 10 although it has a unity execution time.
- Periodic tasks may be interrupted to have aperiodic tasks execute. Aperiodic tasks will have reduced response time but periodic or sporadic tasks may miss deadlines.
- A better method than the above-described procedures called *slack stealing* works similar to background scheduling of aperiodic tasks. The *slack time* (or the laxity) of a periodic time is the time interval between its deadline and its remaining time of computation. If $C_i(t)$ is the remaining computation time of a task τ_i at time t, the slack or the laxity of τ_i at t is,

$$slack_i = d_i - t - C_i(t)$$

In slack stealing, a periodic task may be removed to a further scheduling point without missing its deadline, to leave time for aperiodic task scheduling. Implementation of this method to tasks of Fig. 7.16 is shown in Fig. 7.17. The response of task τ_3 is reduced to 1 using slack stealing in this example. Calculation of slack and moving the slack requires significant computation.

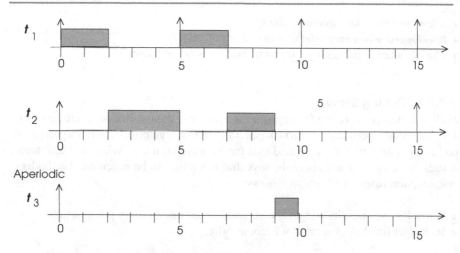

Fig. 7.16 Background scheduling of an aperiodic task

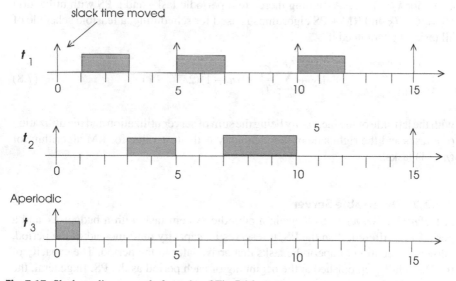

Fig. 7.17 Slack stealing example for tasks of Fig. 7.16

7.7.2 Periodic Servers

A *periodic server* is a system task that is scheduled like any other periodic tasks with the purpose of executing aperiodic tasks. It is specified as $\tau_S = (E_S, T_S)$ where T_S is the period of the server and E_S is its maximum budget. The budget may be *consumed* and *replenished* and is *exhausted* when it finishes. A periodic server can be in one of the following states:

- *idle* if aperiodic task queue is empty.
- *backlogged* when aperiodic task queue is not empty.
- *eligible* when it is backlogged and its budget is not exhausted.

7.7.2.1 Polling Server

Polling Server (PS) is aimed to improve the average response time of aperiodic tasks. It is formed as a periodic system task generally with a high priority, it has a period T_S and computation time C_S allocated to it for each period it has. When it is activated, it uses its budget for any aperiodic task that is waiting to be executed. The budget consumption rules for PS are as follows:

- When PS executes, its budget is consumed at the rate of unity per unit time.
- Its budget finishes as soon as it becomes idle.

The budget of PS is replenished at the beginning of each of its period, at times $k \cdot T_S$ for $k = 1, \ldots$. Assuming there are n periodic tasks and a PS with utilization $U_S = C_S/T_S$ and RM + PS algorithms is used for scheduling, a feasible schedule of all tasks is guaranteed if,

$$U_S + \sum_{i=1}^{n} \frac{C_i}{T_i} \leq (n+1)(2^{\frac{1}{n+1}} - 1) \tag{7.8}$$

with the left side of the inequality being the sum of server utilization and the utilization of n tasks and the right side of the inequality is the test value for RM algorithm for $(n+1)$ tasks.

7.7.2.2 Deferrable Server

A Deferrable Server (DS) is again a periodic system task with a budget E_S and a period T_S. Different than the PS, it reserves its capacity until the end of the period, allowing execution of aperiodic tasks that arrive later in the period. The capacity of the DS is fully replenished at the beginning of each period as the PS. In general, the DS provides better response times for aperiodic tasks.

7.7.2.3 Sporadic Server

The Sporadic Server (SS) is a high-priority task for aperiodic task scheduling and works similar to DS by reserving its budget until the end of its period, waiting for an aperiodic request. However, its replenishment of its capacity is different; SS replenishes its capacity only after it has been consumed by aperiodic task executions. The complexity of the implementation of the SS is higher than implementing the PS and the DS due to hardness of computation of replenishment instants and the timer management.

7.7.2.4 Dynamic-Priority Servers

A dynamic-priority server has a varying priority as its name implies. The following are the main types of dynamic servers:

- *Constant Bandwidth Server* (CBS): This type of server has a known processor time U_{CBS} reserved to it. It has a budget, and consumption and replenishment rules like other servers and it is scheduled with the EDF principle when its budget is not zero. The utilization of this server is constant and hence the name. Its budget and deadline are determined so as to provide it with a fixed utilization and it can have a budget to consume only when it executes. The replenishment rules for CBS are the following:

  - It has a zero budget and a zero deadline initially.
  - When an aperiodic task with an execution time C_A arrives at time t_A.

    If $t_A < D_{CBS}$ wait for CBS to finish
    If $t_A \geq D_{CBS}$, set $D_{CBS} = t_A + C_A / U_{CBS}$ and $C_{CBS} = C_A$

  - When the deadline of the CBS is reached at time t_D, if the CBS is backlogged, set $D_{CBS} = D_{CBS} + C_A / U_{CBS}$ and $C_{CBS} = C_A$. Otherwise, wait since the server is busy.

  This way, it is guaranteed that the CBS always has enough budget to finish the task at the front of the aperiodic task queue when it is replenished.
- *Total Bandwidth Server* (TBS): This server provides a better response time for aperiodic tasks than the CBS by using the time not used by the periodic tasks. It has similar replenishment rules to the CBS with the following difference when an aperiodic task with an execution time C_A arrives at time t_A.

  - set $\max(D_{CBS}, t) + C_A / U_{CBS}$ and $C_{CBS} = C_A$.

  When TBS finishes execution of the current aperiodic task τ_A, this task is removed from the aperiodic task queue and if the CBS is backlogged, set $D_{CBS} = D_{CBS} + C_A / U_{CBS}$ and $C_{CBS} = C_A$. Otherwise, wait since the server is busy. The TBS is always ready for execution when it is backlogged.

7.8 Sporadic Task Scheduling

Sporadic tasks have a minimum interval between their activations but the exact time of their activation is not known beforehand. They have hard deadlines and hence, we may not be able to schedule them. These tasks go through a *acceptance test* by the scheduler and a decision is made whether to run them or not. We need to make sure that when a new sporadic task arrives, accepting it will not cause missing of

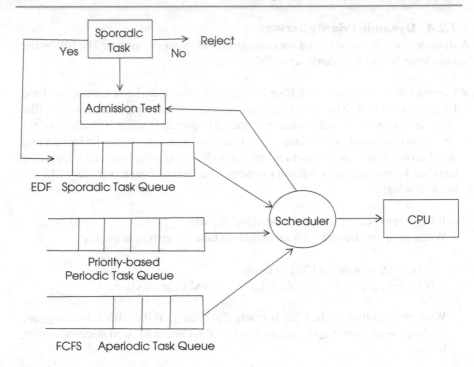

Fig. 7.18 Scheduling of sporadic, periodic and aperiodic tasks

a previously accepted sporadic task. Accepted tasks will be executed using EDF algorithm since this method is convenient in a dynamic-priority- based system. A typical scheduling scenario in a real-time system will look like the one in Fig. 7.18.

A sporadic task is specified as $S_i(A_i, C_i, D_i)$ with A_i as arrival time, C_i is the computation time and D_i is the deadline in the usual manner. The density of a sporadic task is,

$$\Delta_i = \frac{C_i}{D_i - A_i}$$

The total density of a system with n tasks is,

$$\Delta = \sum_{i=1}^{n} \Delta_i$$

Theorem 7.2 *A system of independent preemptable sporadic tasks is schedulable using the EDF algorithm if the total density of all active jobs in the system are less than or equal to 1 at all times.*

The admission test for a sporadic task is based on this theorem. When a sporadic task S_i arrives, if it is determined that the total density will exceed 1 in the feasible interval, S_i is rejected.

7.9 DRTK Implementation

In this section, we will first show how to implement three scheduling policies in DRTK; the Rate Monotonic Scheduler, the Least Laxity First Scheduler for periodic tasks and the Deferrable Server for aperiodic tasks. Note that the scheduler is a POSIX thread that is awakened by a semaphore *post* call of POSIX in DRTK. We will design it as a simple fast function that determines which task to run and signals the POSIX semaphore of that task to make it run.

7.9.1 The Rate Monotonic Scheduler

We will assume tasks are periodic, independent with no precedence constraints and they do not share memory. Also, RM schedulability test is performed offline so that all tasks can meet their deadlines. The RM scheduler of DRTK assumes that the periodic tasks are in a priority queue *RM_Queue* and hence, we need to have a function that queues the tasks in the initialization of DRTK. This system call named *Init_RM* shown below calculates priorities of periodic tasks and inserts them in *RM_Queue*. It should be called in the initialization procedure *init_system* of DRTK.

```
/*******************************************************************
                    Initialize RM tasks
********************************************************************/

void Init_RM() {
    task_ptr_t task_pt;
    for (i=0; i<N_TASKS; i++)
      if( task_tab[i].type=PERIODIC) {
          task_tab[i].priority=(int)1/task_tab[i].period;
          task_pt=&(task_tab[i]);
          insert_queue(&RM_queue,task_pt);
      }
}
```

Let us review the approach in DRTK for periodic tasks. An RM task in DRTK delays itself for a time interval by the *delay_task* system call and waits to be awakened. The *Time_ISR* when waken by the timer interrupt, checks *delta_queue* and if the front of the queue has zero time left, it dequeues the task from the front, makes it *ready* and calls the scheduler. This operation is in line with the preemptive RM scheduling. Note that RM scheduling makes dynamic decisions at runtime using static priorities and is preemptive.

The code for RM Scheduler, *RM_Scheduler*, can be formed based on DRTK delay operation as a simple procedure that always waits at its POSIX semaphore as the general scheduling in DRTK. It may have been called either as a result of a timer interrupt since a periodic task needs to be activated or the current task has finished processing and delays itself for the next period. In order to distinguish between these two cases; the *RM_Scheduler* first checks the state of the current task, if it is *delayed*, the first task from the ready queue, *RM_Queue*, is selected. Otherwise, it compares the priority of the current task with the priority of the task in front of *RM_Queue*, and selects the task with the higher priority as the next task to run as shown below. If the current task that has called the scheduler has a higher priority than the first task of *RM_Queue*, it is resumed by activating the task identifier stored in system table as *preempted_tid* by the *Schedule* system call. We need to store the identifier of the preempted task saved since the scheduler task runs as a task of DRTK and hence as a POSIX thread. We assume there is at least one *ready* task in *RM_Queue*.

```
/*************************************************************
                    RM Scheduler
*************************************************************/
TASK RM_Scheduler(){

 task_ptr_t task_pt;
 while(TRUE){
  sem_wait(&(System_Tab.sched_sem));
  if(task_tab[current_tid].state!=DELAYED ||
       task_tab[current_tid].state!=BLOCKED)
   if(task_tab[current_pid].priority <
               (RM_Queue.front)->priority) {
    task_pt=dequeue_task(&RM_queue);
    current_tid=task_pt->tid;
    insert_task(&RM_Queue,&task_tab[System_Tab.preempted_tid]);
    }
   else
     current_tid=System_tab.preempted_tid;
  else {
    task_pt=dequeue_task(&RM_queue);
    current_tid=task_pt->tid;
  }
  current_pt=&(task_tab[current_pid]);
  sem_post(&(task_tab[current_tid].sched_sem));
 }
}
```

7.9.2 Earliest Deadline First Scheduler

The EDF scheduler should always run the task with the closest deadline. Preemption occurs when a task with a deadline closer to that one of the running is activated. We will assume all tasks are periodic and independent and they do not share resources. Periodic tasks are inserted in the delayed task queue *delta_queue* and the clock interrupt service routine decrements the delay value of the first task in the queue. If this value is zero, task is dequeued and made ready in the usual operation of the DRTK. The EDF scheduler is a system task invoked when a delayed task from *delta_queue* is activated or the current task is blocked waiting for an event. All the *ready* tasks are in *EDF_Queue* and the EDF Scheduler compares the deadline of the running task with the deadline of the unblocked task in *EDF_Queue* and selects the one with the nearest deadline as shown in the code below. We assume absolute deadlines are stored in the task control block data structures of tasks. We will assume there is at least one *ready* task in *EDF_Queue* as in RM algorithm.

```
/*****************************************************************
                        EDF Scheduler
*****************************************************************/
TASK EDF_Scheduler(){

 task_ptr_t task_pt;
 while(TRUE) {
  sem_wait(&(System_Tab.sched_sem));
   if(task_tab[current_tid].state!=DELAYED ||
          task_tab[current_tid].state!=BLOCKED)
    if(task_tab[current_pid].abs_deadline >
          (EDF_Queue.front)->abs_deadline) {
     task_pt=dequeue_task(&EDF_Queue);
     current_tid=task_pt->tid;
     insert_task(&EDF_Queue,&task_tab[System_Tab.preempted_tid]);
    }
   else
      current_tid=System_tab.preempted_tid;
   else {
      task_pt=dequeue_task(&EDF_queue);
      current_tid=task_pt->tid;
   }
   current_pt=&(task_tab[current_pid]);
   sem_post(&(task_tab[current_tid].sched_sem));
  }
}
```

7.9.3   The Least Laxity First Scheduler

We assume tasks are periodic with hard deadlines, they do not have any precedence relationships and they do not share resources as before for this algorithm. The *laxity* of a task is the interval between its deadline and the time it finishes. The least laxity first (LLF) Scheduler, *LLF_Scheduler*, is also priority based. This time, however, we need to test laxities of tasks at runtime at scheduling points. We will assume that initial laxities of tasks are calculated and stored at system initialization with the following routine by subtracting the relative deadline value from the worst-case execution time (WCET) of a task. These tasks are made *ready* and inserted into *LLF_Queue* with respect to their laxity times.

```
/*****************************************************************
                     Initialize LLF tasks
*****************************************************************/
void Init_LLF() {

  task_ptr_t task_pt;
```

```
for (i=0; i<N_TASKS; i++)
  if( task_tab[i].type=PERIODIC) {
    task_tab[i].laxity=task_tab[i].rel_deadline-task_tab[i].wcet;
    task_pt=&(task_tab[i]);
    insert_task(&LLF_queue,task_pt);
  }
}
```

The activation of *LLF_Scheduler* is done either when a task finishes execution or a new instance of a periodic task needs to be activated when its period starts. If the current task is in *delayed* or *blocked* state, the scheduler simply dequeues the first task from *LLF_Queue* and dispatches it, otherwise, it compares laxity value of the current task with the laxity value of the first task on *LLF_Queue* and selects the one with less laxity to run as in the code below. Note that we need to add the execution time field to the task control block to be able to calculate its current laxity online. Calculation of the executed time of a task may be updated at each clock tick by adding the time quanta value to the value in task control block, which is not implemented. Our assumption is that there exists at least one task in the *LLF_Queue*.

```
/********************************************************************
                         LLF Scheduler
*********************************************************************/
void LLF_Scheduler() {

 task_ptr_t task_pt;
 while(TRUE) {
  sem_wait(&(System_Tab.sched_sem));
  if(task_tab[current_tid].state!=DELAYED) ||
      (task_tab[current_tid].state!=BLOCKED) {
    task_tab[current_tid].laxity=
    task_tab[current_tid].rel_deadline -
             task_tab[current_tid].executed;
    if(task_tab[current_pid].laxity >
                  (LLF_Queue.front)->laxity) {
     task_pt=dequeue_task(&LLF_queue);
     current_tid=task_pt->tid;
     insert_task(&LLF_Queue,&task_tab[System_Tab.preempted_tid]);
    }
    else
      current_tid=System_Tab.preempted_tid;
  }
  else {
    task_pt=dequeue_task(&LLF_queue);
    current_tid=task_pt->tid;
  }
  current_pt=&(task_tab[current_pid]);
  sem_post(&(task_tab[current_tid].sched_sem));
  }
}
```

7.9.4   Polling Server

The polling server (PS) is a high-priority system task that is scheduled like the rest of real-time tasks. When scheduled, it checks the aperiodic task queue *AT_Queue* and if there is an aperiodic task waiting, it is dequeued and made ready by this server.

```
/*****************************************************************
                        Polling Server
*****************************************************************/
TASK Polling_Server()
   task_ptr_t task_pt;

   while(TRUE){
      delay_task(current_tid);
      if(AT_Queue.front!=NULL)
        { task_pt=dequeue_task(&AT_Queue);
          unblock_task(task_pt->tid, YES);
        }
    }
}
```

7.10   Review Questions

1. What is the difference between the absolute deadline and the relative deadline of a task?
2. What is the utilization of a periodic real-time task?
3. What is meant by preemptive and non-preemptive scheduling of tasks?
4. What makes a task set dependent?
5. What is the difference between table-driven scheduling and the cyclic executive?
6. How can dynamic scheduling of static priority tasks be achieved?
7. What is the main principle of RM scheduling?
8. What is the main principle of EDF scheduling?
9. Can EDF scheduling be used for aperiodic tasks?
10. What is the difference between EDF scheduling and LLF scheduling?
11. What is the response time of a task and how is it affected by other tasks?
12. How does a Polling Server work?
13. What is the difference between a Polling Server and a Deferrable Server?
14. What is the main idea of a Sporadic Server?
15. How does a Constant Bandwidth Server work?

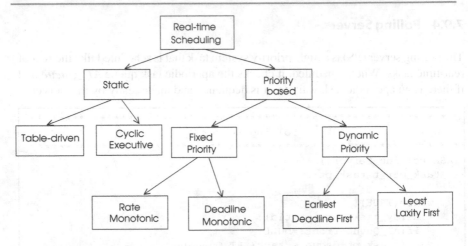

Fig. 7.19 Detailed classification of uniprocessor-independent task scheduling

7.11 Chapter Notes

We reviewed basic scheduling algorithms in a single processor for independent tasks that do not share resources. We have different strategies when preemption is allowed or not as we saw. In the cyclic executive model, an offline schedule of periodic real-time tasks are computed using the lowest common multiple of their periods and the scheduling is performed by obeying the table prepared beforehand while running tasks. RM scheduling algorithm is commonly used for hard real-time periodic tasks that are independent and do not share resources. This method assigns fixed priorities to tasks offline based on their periods. It is simple to implement, however, has a strict upper bound on processor utilization. EDF algorithm works with dynamic priorities and the priority of a task reflects its closeness to its deadline. This algorithm ensures that the highest priority task is running at all times and obtains better utilization of the processor. However, EDF algorithm requires evaluation of dynamic priorities of tasks based on their closeness to their deadlines and hence has significant runtime overhead. For this reason, RM algorithm is more frequently preferred to the EDF algorithm. The LLF algorithm basically works on a similar principle to the EDF algorithm but considers computation times of tasks during scheduling. It also suffers from overheads as the EDF algorithm since dynamic priorities are to be computed at runtime. We can now form a detailed taxonomy of uniprocessor-independent task scheduling algorithms based on our analysis as shown in Fig. 7.19.

There are significantly different methods for aperiodic and sporadic tasks. Aperiodic tasks can be scheduled using various server types. For fixed priority systems, the *Polling Server* is the simplest one that has a budget available at predefined scheduling points. The Deferrable Server has also a fixed budget but periodical replenishment is possible. A simple approach to deal with sporadic tasks is to generate a superficial

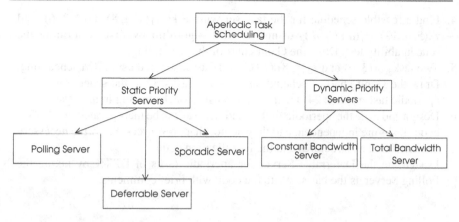

Fig. 7.20 Classification of aperiodic servers

task and if a sporadic task is available at this allocated time, execute it. Otherwise, this allocated time frame may be used for periodic tasks. The *Sporadic Server* serves sporadic tasks and has a fixed budget that is replenished if it was consumed. The dynamic-priority systems have the *Constant Bandwidth Server* and the *Total Bandwidth Server* in which deadlines are independent on execution time in the former and dependent in the latter. The classification of aperiodic server algorithms is depicted in Fig. 7.20.

Finally, we showed a sample set of scheduling algorithms in DRTK; the RM Scheduler, the LLF Scheduler and the Polling Server with necessary modifications to data structures and the initialization routines.

7.12 Exercises

1. A periodic independent task set with tasks $\tau_1(1, 6)$, $\tau_2(3, 8)$, and $\tau_3(2, 12)$ $(\tau_i(C_i, T_i))$ is given.

    a. Work out the scheduling table for this set using table-driven scheduling and draw the Gantt charts of task executions.
    b. Find the hyperperiod H and frame length f to run Cyclic Executive for these tasks and draw the Gantt charts of task executions.

2. Modify the Cyclic Executive pseudocode of Algorithm 7.2 so that aperiodic tasks in an FCFS queue may be executed if there is a slack time left before the next timer interrupt occurs.

3. Find a feasible schedule for the task set with tasks $\tau_1(3, 8)$, $\tau_3(2, 12)$, and $\tau_3(4, 36)$ $(\tau_i(C_i, T_i))$ using the RM algorithm by first performing the schedulability test. Draw the Gantt chart of the scheduling.

4. Find a feasible schedule for the task set with tasks $\tau_1(1, 3, 8)$, $\tau_3(2, 3, 6)$, and $\tau_3(5, 4, 12)$ $(\tau_i(a_i, C_i, d_i))$ using the EDF algorithm by first performing the schedulability test. Draw the Gantt chart of the scheduling.
5. Two tasks $\tau_1(3, 6)$ and $\tau_3(3, 8)$ $(\tau_i(C_i, T_i))$ are scheduled using RM scheduling. Draw the Gantt chart of scheduling for these tasks. Find the scheduling of an aperiodic task $\tau_3(2, 4, 24)$ $(\tau_i(a_i, C_i, d_i))$ using slack stealing in this system.
6. Design and code the aperiodic EDF scheduler that can be incorporated into DRTK code. Tasks are independent and they do not share resources. Show the necessary additions to data structures and the initialization code.
7. Design the the Deferrable Server for aperiodic tasks of DRTK by taking the Polling Server as the basis. Write the code with brief comments.

References

1. Buttazzo (1993) Hard real-time computing systems: predictable scheduling algorithms and applications. Real-time systems series, 3rd edn. Springer
2. Liu CL (2000) Real-time systems. Prentice Hall
3. Liu CL, Layland JW (1973) Scheduling algorithms for multiprogramming in a hard-real-time environment. J ACM 20(1):40–61
4. Mullender S (1993) Distributed systems, 2nd edn. Addison-Wesley

Uniprocessor-Dependent Task Scheduling

8

8.1 Introduction

We have reviewed uniprocessor algorithms assuming tasks are independent without any interactions. In many real-life applications, tasks exchange data and there is a precedence relationship between tasks. Moreover, tasks may share resources which are accessed by mutual exclusion of tasks. These precedence relations and sharing of resources affect the scheduling decisions significantly.

In the first part of this chapter, we review basic algorithms for aperiodic tasks with deadlines and with precedence relationship between them but they do not share resources. These tasks are conveniently represented by a directed acyclic graph (DAG) and the algorithms presented use modified forms of the basic algorithms reviewed in the previous chapter to achieve meeting of deadlines of all tasks. Two algorithms presented are the latest deadline first (LDF) and the modified earliest deadline first (MEDF) algorithms.

Priority inversion happens when a lower priority task holds a resource and blocks a higher priority task from executing. We revise protocols to solve the priority inversion problem in the second part of the chapter. Two protocols analyzed are the Priority Inheritance and the Priority Ceiling protocols. Finally, we implement the LDF, MEDF algorithms and the Priority Inheritance protocols in DRTK.

8.2 Dependent Task Scheduling

In our first attempt, we will assume the following in search of a solution to the independent task scheduling problem; tasks do not share resources but there is a precedence relationship between them. This problem is NP-hard in the general case

© Springer Nature Switzerland AG 2019
K. Erciyes, *Distributed Real-Time Systems*, Computer Communications and Networks, https://doi.org/10.1007/978-3-030-22570-4_8

and we will review two polynomial heuristic algorithms for this purpose; latest deadline first algorithm and the modified earliest deadline first algorithm.

8.2.1 Latest Deadline First Algorithm

The principle of the latest deadline first (LDF) algorithm is to delay scheduling of a task that has the latest deadline as long as it has no successors. Having one or more successors of a task τ_i means delaying τ_i causes delays of its successors and therefore should be avoided. Hence, the algorithm searches tasks that have no successors and selects the one with the latest deadline among such tasks. The selected task is pushed onto a stack and the number of successors of each of its predecessors is decremented. Continuing in this manner, all tasks are put on top of the stack and then each task is popped of the stack and scheduled as shown in Algorithm 8.1. This algorithm starts selecting one of the terminal vertices of the DAG in practice. Note that the task with the latest deadline is processed first for scheduling but scheduled later than unselected tasks.

Algorithm 8.1 *Latest Deadline First*

1: **Input**: set $T = \{\tau_1, \ldots, \tau_n\}$ of n aperiodic tasks, their precedence set $S = \{s_1, \ldots, s_k\}$
2: **Output**: Schedule table for tasks
3: S: stack of tasks
4: **for** i=1 to n **do**
5:     **select** task τ_x that has no successor and has the latest deadline
6:     **push** τ_x onto S
7:     **decrement** number of successors of each predecessor of τ_x
8: **end for**
9: **for** i=1 to n **do**
10:     **pop** τ_x from S
11:     **enqueue** τ_x in scheduling queue
12: **end for**

Selecting the task with no predecessor requires $O(n)$ operations in the simple application of this algorithm, totalling $O(n^2)$ operations. A dependent task set with seven tasks, $\tau_1, \ldots, \tau_7$, 0 arrival times and absolute deadlines are shown in Fig. 8.1.

Pushing these tasks onto a stack and scheduling them in reverse order results in the Gantt chart of Fig. 8.2. We can see all tasks meet their deadlines for this set without needing any preemption.

8.2.2 Modified Earliest Deadline First Algorithm

The EDF algorithm proved to be optimal in one processor and independent tasks case and hence, we will attempt to apply this algorithm for the dependent task scheduling

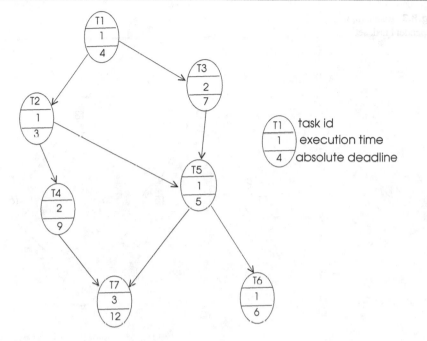

Fig. 8.1 The task graph of a dependent task set

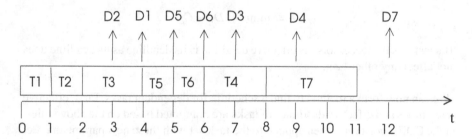

Fig. 8.2 LDF scheduling of tasks of Fig. 8.1

problem. We assume tasks do not share resources but there is a precedence relationship between them. We need to modify the release times and deadlines of tasks
according to precedences as follows assuming $\tau_i(a_i, C_i, D_i) \prec \tau_j(a_j, C_j, D_j)$:

- Task τ_j must start execution later than its release time and later than the maximum
  finishing time of its predecessors, therefore,

$$a'_j = \max(a_j, a_i + C_i)$$

If a task has no predecessors, its effective release time is its release time since it
does not have to wait for any task to start execution.

Fig. 8.3 An example
dependent task set

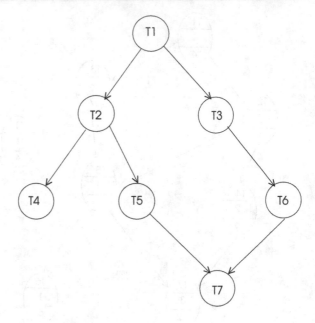

- A task τ_i must finish before its deadline and,

$$D_i' = \min(D_i, D_j' - C_j)$$

If a task has no successors, its effective deadline is its deadline as its deadline does
not affect any other task.

We can now form a method to find a schedule for dependent tasks. First, effective
release times and effective deadlines of tasks are computed based on the above rules,
and the EDF algorithm is then applied to the task set with these new parameters. Let
us consider the task graph of Fig. 8.3.

Table 8.1 An example task set

τ_i	C_i	D_i	a_i'	D_i'
1	1	3	0	3
2	1	5	1	5
3	1	4	1	4
4	1	7	2	7
5	1	5	2	5
6	1	7	2	5
7	1	6	3	6

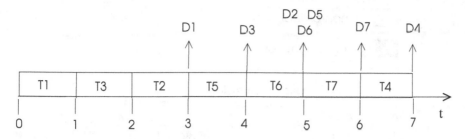

Fig. 8.4 MEDF scheduling of tasks of Fig. 8.3

The characteristics of these tasks are depicted in Table 8.1; each task has a unity execution time and all are available at time 0. The last two columns of the table show the modified arrival times and modified deadlines.

The MEDF scheduling of these tasks is depicted in Fig. 8.4. Note that we could have preempted a running task at time t if a task with a closer deadline at t becomes ready.

8.3 Scheduling of Resource Sharing Tasks

Real-time tasks share resources such as data structures, files, input/output units, or main memory in real applications. These resources must be protected against simultaneous accesses by mutual exclusion mechanisms provided by the operating system. The segment of code that a task accesses a shared resource is called a critical section and semaphores are commonly used to provide mutual exclusion as we have reviewed in Chap. 4. Typically, a task that wants to start executing its critical section performs a *wait* on the semaphore that protects the resource which may cause it to be blocked if another task is using the resource, and when it finishes executing its critical section in which it accesses the resource, it issues a *signal* call on the semaphore to release any waiting tasks on the semaphore. Tasks that wait for the same resource are queued on the semaphore that protects the resource.

Most real-time scheduling methods are priority based, that is, each task has a priority and the task with the highest priority is scheduled at any time. Priority of a task may be static which does not change with time as in Rate Monotonic scheduling or dynamic which is assigned with regards to deadlines of tasks as in Earliest Deadline First method. Consider the case with two tasks τ_1 and τ_2 as shown in Fig. 8.5 with lower task identifier denoting a higher priority. Let us assume τ_2 enters its critical section to access resource R at time t_1 and then is preempted by τ_1 at time t_2 since τ_1 has a higher priority. Next, τ_1 wants to access R and since τ_2 holds R, it is blocked by the operating system on the semaphore that protects R at time t_3. Task τ_1 has to wait until τ_2 finishes with the resource and releases it by signaling the semaphore at time t_4 after which τ_1 enters its critical section and accesses R. It finishes critical section

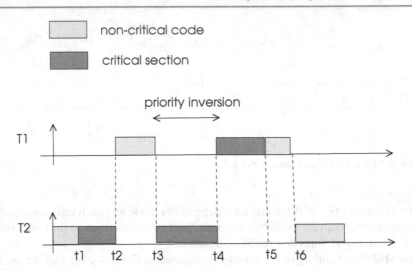

Fig. 8.5 Priority inversion with two tasks

at time t_5 and then finishes its noncritical code at time t_6 at which time τ_2 starts executing its noncritical code. The sequence of described events has led to a higher priority task waiting for a lower priority task which is called *priority inversion*. The waiting time of τ_1 is bounded by the time it takes τ_2 to finish its critical section and hence, this type of priority inversion is called *bounded* priority inversion.

A more serious case is when there are three tasks τ_1, τ_2, and τ_3 in the order of decreasing priority in the system as depicted in Fig. 8.6.

The following sequence of events occurs in this figure:

1. Task τ_3 which has the lowest priority executes a wait on semaphore S and enters its critical section to access shared resource R at time t_1.
2. Task τ_3 is preempted by task τ_1 since $P(\tau_1) > P(\tau_3)$ at time t_2.
3. Task τ_1 waits on S since τ_3 is blocked on this semaphore at time t_3.
4. The medium priority task τ_2 has a higher priority than τ_3 and since τ_1 is blocked, it is scheduled to run until it finishes at time t_4.
5. Task τ_3 is the only runnable task and hence runs until it finishes its critical section at time t_5.
6. Task τ_1 is signaled now by τ_3 on semaphore S and starts and runs its critical section until time t_6 and then runs its noncritical code until time t_7 and finishes.
7. Task τ_3 is scheduled again as it is the only remaining task in the system and runs to completion until time t_8.

Different than the previous case, task τ_2 may run for a long time causing the high-priority task τ_1 to miss its deadline. This situation is called *unbounded waiting* case of priority inversion. A possible solution to this problem is to disable preemption when a task is in its critical section, however, many unrelated tasks that do not share

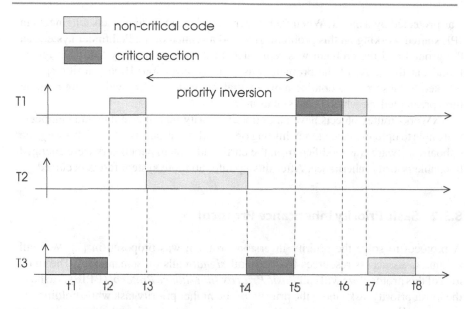

Fig. 8.6 Priority inversion with three tasks

the resource that is blocked will have to wait possibly causing missing of deadlines. Two efficient solutions to this problem are the priority inheritance protocol and the priority ceiling protocol. We will first review the case of Mars Pathfinder.

8.3.1   The Case of Mars Pathfinder

Mars Pathfinder was designed and built by the Jet Propulsion Laboratory (JPL) of California Institute of Technology to land on Mars to collect scientific data about Mars [1]. Pathfinder was accommodated with the VxWorks real-time operating system which provides preemptive fixed priority scheduling in the form of a cyclic scheduler. The spacecraft started resetting itself soon after it landed on Mars on July 4, 1997, which resulted in delay of data collection although no data was lost. Pathfinder had three main periodic tasks:

- τ_1: Information Bus Thread; acts as the bus manager, has high frequency, high priority.
- τ_2: Communication Thread; has medium frequency, and medium priority with long execution time.
- τ_3: Weather Thread; collects geological data, has low frequency and low priority.

Each task checked if others executed in the previous cycle and if the test fails, system is reset by a watchdog timer. A shared memory was used for communication between the tasks. Data was passed from τ_3 to τ_2 over τ_1 and access to shared memory was protected by a *mutex*. When Pathfinder experienced system resets, engineers in JPL started working on this problem. JPL had a replica of the Pathfinder system on the ground and the problem was replicated in the lab after many tests. Designers found out the cause of the problem was priority inversion. High priority τ_1 was blocked because τ_3 was holding a mutex that τ_1 needed and τ_2 with a long running time preempted τ_3, which is the situation in Fig. 8.6.

VxWorks mutex objects have an optional priority inheritance flag and engineers were able to upload a patch to set this flag on the information bus mutex. This way, the onboard software was modified from the earth and system parameters were changed to enable priority inheritance. After this update, no more system resets occurred.

8.3.2 Basic Priority Inheritance Protocol

A protocol to solve the priority inversion problem was proposed in [2]. We will assume tasks access resources by *wait* and *signal* calls on semaphores. The main idea of this protocol we will call *Basic Priority Inheritance Protocol* (BPIP) is to have the lower priority task inherit the priority of the higher priority task while holding the resource. We further assume tasks have priorities and are scheduled based on their priorities, that is, the highest priority task runs at any time. The following shows a possible sequence of events in this protocol:

1. A task τ_i issues a *wait* call on the semaphore S associated with the resource R and gets blocked since R is held by a lower priority task τ_j.
2. Task τ_i transfers its priority P_i to τ_j when this happens, hence $P_j = P_i$.
3. When task τ_j finishes executing its critical section, it transfers its current priority to τ_i, hence $P_i = P_j$ and it restores its original priority.

The running of tasks of Fig. 8.6 is now as shown in Fig. 8.7 using the priority inheritance protocol. We can see task τ_3 inherits the priority of task τ_1 in between times t_3 and t_4 and can finish its critical section. Task τ_1 has its priority back at t_4 and can enter its critical section. Note that running of the intermediate priority task τ_2 is now delayed until time t_5 and only at time t_6 when τ_2 finishes executing its noncritical code, τ_3 can start running its noncritical code.

Priority inheritance is transitive, that is, if a task τ_i is blocked by τ_j which is blocked by τ_k with $P_i > P_j > P_k$, then τ_k inherits the priority of τ_i through τ_j. Analysis of the BPIP shows two main problems which are the long delays and deadlocks as described below.

Long Delays

Assume n tasks $\tau_1, \tau_2, \ldots, \tau_n$ in the order of decreasing priorities. Consider the case where τ_n starts a critical section after a *wait* on semaphore S after which τ_{n-1} is scheduled which performs a *wait* on S and gets blocked, and passes its priority to τ_{n-1}. Let us further assume this situation continues up to task τ_1 priority of which is passed all through other tasks to task τ_n. The highest priority task τ_1 has to wait

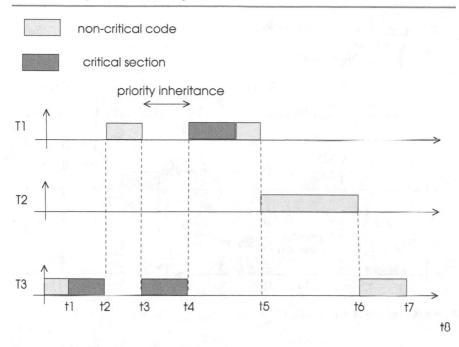

Fig. 8.7 Priority inheritance example

execution of n critical sections which may be long for a large value of n enough to cause τ_1 miss its deadline. This situation ic called *chain blocking* as depicted in Fig. 8.8.

There are two semaphores S_1 and S_2 in the system and tasks τ_1, τ_2 and τ_3 have priorities inversely proportional to their indices. The following are the sequence of events in the figure:

1. t_1: Task τ_3, which is the lowest priority task, performs a *wait* on S_1 and locks it and starts executing its critical section.
2. t_2: Task τ_2 preempts τ_3 as its has a higher priority and starts its noncritical section.
3. t_3: Task τ_2 performs a *wait* on S_1 and locks it and starts executing its critical section.
4. t_4: Task τ_1 preempts τ_2 as its has a higher priority and starts its noncritical section.
5. t_5: Task τ_1 issues a *wait* on S_1 and since this semaphore is locked by τ_1, it transfers its priority to τ_3 which continues executing its critical section.
6. t_5: Task τ_3 finishes executing its critical section, signals S_1 and restores its priority and the priority of τ_1.
7. t_6: Task τ_1 can now enter its critical section protected by semaphore S_1.
8. t_7: Task τ_1 exits its critical section, signals S_1 and starts its noncritical section.
9. t_8: Task τ_1 performs a wait on semaphore S_2 and since S_2 is locked by task τ_2, it transfers its priority to τ_2 which starts executing its critical section.

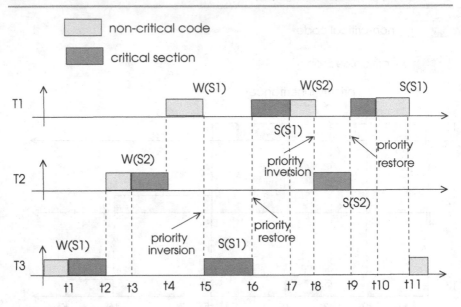

Fig. 8.8 An example of chain blocking

10. t_8: Task τ_1 performs a wait on semaphore S_2 and since S_2 is locked by task τ_2, it transfers its priority to τ_2 which starts executing its critical section.
11. t_9: Task τ_1 is scheduled as performs a wait on semaphore S_2 and since S_2 is locked by task τ_2, it transfers its priority to τ_2 which starts executing its critical section.
12. t_{10}: Task τ_1 finishes its critical section protected by S_2 and signals S_2. It then continues executing its noncritical section.
13. t_{11}: Task τ_1 finishes its noncritical section at which time task τ_3 starts its noncritical section and then finishes it.

This example demonstrated that the highest priority task τ_1 out of three tasks had to wait for τ_3 for semaphore S_1 and τ_2 for semaphore S_2. In general, the chain blocking may cause a task τ_i to wait for a blocking time B_i,

$$B_i = \sum_{k=1}^{M} block(k, i)C(k) \tag{8.1}$$

time where M is the number of critical sections of lower priority tasks, $block(k, i)$, $C(k)$ is the is the worst-case execution time of critical section k. The utilization of the processor including the blocking time for the RM scheduling can be modified to consider blocking times of tasks as follows:

$$\sum_{i=1}^{n} \frac{C_i}{T_i} + \frac{B_i}{T_i} = U_i + \frac{B_i}{T_i} \leq n(2^{1/n} - 1) \tag{8.2}$$

Fig. 8.9 An example of deadlock caused by BPIP

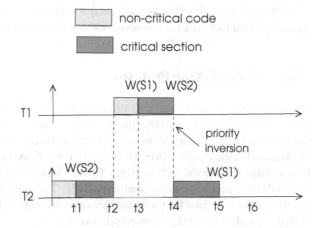

where B_i is the longest priority inversion time that τ_i may wait. In fact, $B_i = CS_j + CS_1 + \cdots + CS_k$ where CS values are the critical sections of lower priority tasks that could block τ_i. The response time of a task τ_i is affected by the blocking time B_i as follows:

$$R_i = C_i + B_i + \sum_{j=1}^{i-1} \left\lceil \frac{R_i}{T_j} \right\rceil C_j \le D_i \qquad (8.3)$$

Deadlock possibility

Consider the case of two tasks τ_1 and τ_2 which issue wait and signal calls to two semaphores S_1 and S_2 as follows:

$$\tau_i : \{wait(S_1), \ldots, wait(S_2), signal(S_2), signal(S_1)\}$$

$$\tau_j : \{wait(S_2), \ldots, wait(S_1), signal(S_1), signal(S_2)\}$$

This scenario may lead to the deadlock depicted in Fig. 8.9.

The sequence of events in this figure are as follows:

1. t_1: Task τ_2 performs a *wait* on S_2 and locks it and starts executing its critical section.
2. t_2: Task τ_1 preempts τ_2 as its has a higher priority and starts its noncritical section.
3. t_3: Task τ_1 performs a *wait* on S_1 and locks it and starts executing its critical section.
4. t_4: Task τ_1 issues a wait on semaphore S_2 inside its critical section protected by S_1 and since τ_2 holds S_2, it passes its priority to τ_2 which continues executing its critical section protected by S_2.
5. t_4: Task τ_2 issues a wait on semaphore S_1 inside its critical section protected by S_2 and since τ_2 holds S_1, it cannot proceed.

In this situation, τ_1 holds S_2 and needs to unlock S_1 to continue and τ_2 holds S_1 and needs to unlock S_1 to continue which is a deadlock condition.

8.3.3 Priority Ceiling Protocol

Priority ceiling protocols solve all of the problems of BPIP. The main idea in these protocols is to anticipate the blocking of a task. If a task can potentially block a higher priority task by locking a semaphore, it is not allowed to do so. These protocols can be classified as the Original Priority Ceiling Protocol and Immediate Priority Ceiling Protocol described in the next sections. The original priority ceiling protocol (OPCP) solves the two problems associated with BPIP. The main idea of this protocol is to associate a *ceiling value* for each semaphore in the system that is equal to the highest priority of tasks that use the semaphore. Formally,

$$ceil(S_i) = \max\{Prio_i | \tau_i \text{ that uses } S_k\}$$

This value is static and can be computed offline. The working of OPCP is based on the following rules:

- The ceiling of a semaphore, $ceil(S)$ is the priority of the highest priority task that uses the semaphore at runtime.
- A task τ_i can only lock (perform a wait without getting blocked) a semaphore if $Prio(i)$ is strictly higher than the ceilings of all semaphores currently locked by other tasks.
- Otherwise τ_i is blocked on S, and the task that currently holds S inherits the priority of task τ_i.

The following illustrates the operation of OPCP in detail:

- Let S_m be the semaphore that has the greatest ceiling value among all semaphores in the system.
- **If** $Prio(\tau_i) > ceil(S_m)$ **then** τ_i enters its critical section.
- **Else** τ_i transfers its priority to the task that currently holds as in BPIP.

An example of OPCP with three tasks τ_1, τ_2 and τ_3 in the order of decreasing priorities that use two semaphores S_1 and S_2 is shown in Fig. 8.10. The semaphore ceiling values are, $ceil(S_1) = Prio(\tau_1)$ and $ceil(S_2) = Prio(\tau_1)$.

The following are the sequence of events in this example:

1. t_1: Task τ_3, which is the lowest priority task, performs a *wait* on S_1 and locks it and starts executing its critical section.
2. t_2: Task τ_2 preempts τ_3 as its has a higher priority and starts its noncritical section.

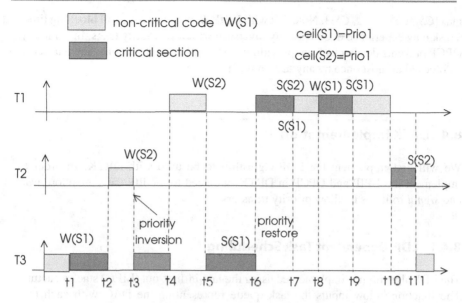

Fig. 8.10 An example of chain blocking

3. t_3: Task τ_2 performs a *wait* on S_2 and is blocked since $Prio_2 < ceil(S_2)$ and Task τ_3 is scheduled since there are no other ready tasks and continues executing its critical section.

4. t_4: Task τ_1 preempts τ_3 and starts its noncritical section.

5. t_5: Task τ_1 issues a *wait* on S_1 and since $Prio_1 < ceil(S_2)$, it is blocked so that τ_3 is started which executes the remaining part of its critical section.

6. t_6: Task τ_3 finishes its critical section and τ_1 is now scheduled which can start its critical section associated with semaphore S_2.

7. t_7: Task τ_1 finishes its critical section, signals semaphore S_2 starts its noncritical section.

8. t_8: Task τ_1 performs a wait on semaphore S_2 and since S_2 is locked by task τ_2, it transfers its priority to τ_2 which starts executing its critical section.

9. t_9: Task τ_1 is scheduled as performs a wait on semaphore S_2 and since S_2 is locked by task τ_2, it transfers its priority to τ_2 which starts executing its critical section.

10. t_{10}: Task τ_1 finishes its critical section protected by S_2 and signals S_2. It then continues executing its noncritical section.

11. t_{11}: Task τ_1 finishes its noncritical section at which time task τ_3 starts its non-critical section and then finishes it.

Analysis

The OPCP is deadlock free and a task τ_i may get blocked at most the duration of the longest critical section. The utilization of the system under RM scheme is the same as the BPIP protocol but the blocking time of a task τ_i is $B_i =$

$\max\{CS_j, CS_{+1}, \ldots, CS_k\}$. Note that with BPIP, B_i was the sum of all blocking times a task may experience due to priority inversion with low-priority tasks. In summary, OPCP prevents deadlocks from providing nesting of critical sections, and a task can be blocked at most once by any task in this protocol.

8.4   DRTK Implementation

We will first implement the LDF algorithm to be used with DRTK. In order to implement the BPIP and OPCP in DRTK, we need to modify the semaphore *wait* and *signal* routines to allow priority transfers.

8.4.1   LDF-Dependent Task Scheduling

The LDF algorithm is implemented using the last-in-first-out (LIFO) stack structure. The function below inputs the task queue representing the DAG with each task containing its predecessors and successors in its task control block. This queue is traversed until a task with no successors is found and in case of more than one such tasks exist, one with the latest deadline is selected and pushed onto the stack (*stack_queue*). A temporary task control block pointer is used to hold the address of the task that has no successors and with the maximum deadline value at each iteration. After pushing the task onto the stack, all of its predecessor tasks' successor counts are decremented. This process continues until all tasks are pushed onto the stack after which, tasks are popped from the stack and enqueued to the output scheduler queue (*sched_queue*).

```
/******************************************************************
           Latest Deadline First Dependent Algorithm
 ******************************************************************/

void LDF( task_queue_t input_task_que ){

    task_queue_t sched_queue, stack_queue;
    task_ptr_t task_pt, temp_pt;
    task_control_block_t task_temp;

    task_pt=input_task_que.front;
    temp_pt=task_pt;
    n_tasks=input_task_que.n_tasks;

    for(i=1;i<=input_task_que.n_tasks;i++){
        while(task_pt->next != NULL){
            if (task_pt->n_successors==0 &&
                task_pt->deadline>temp_pt.deadline)
                temp_pt=task_pt;
            task_pt=task_pt->next;
```

```
        }
        push_stack(stack_queue,task_pt);
        for(j=1;j<=task_pt->n_predecessors;j++){
            task_id=task_pt->predecessors[i];
            temp_pt=&(task_tab[task_id]);
            temp_pt->n_successors--;
        }
    }
    for(i=1;i<=input_taskque_pt->n_tasks;i++){
        task_pt=pop_stack(stack_queue);
        enqueue(sched_queue,task_pt);
    }
    return(DONE);
}
```

8.4.2 Priority Inheritance Protocol

In order to implement BPIP, a task is allowed to have two priorities, a *nominal priority*
and an *active priority*. The nominal priority is the field reserved in task control block
as the priority of the task. The active priority, however, is the current priority of a
task when using BPIP. This new field *active_priority* is added to the task control
data structure. We will modify the *wait* and *signal* system calls of DRTK to enable
BPIP operation. The semaphore *wait* call in DRTK decrements semaphore value and
if this value is negative, meaning another task has already acquired the semaphore,
it gets blocked in the semaphore queue and scheduler is called. We will first modify
the semaphore structure of DRTS by adding the field *holder* that shows the task that
currently holds the semaphore as below. This field is needed to be able to compare
priorities of the task that issues a *wait* and a task that is in its critical section. The
other new field is the *transfer_id* to keep the identifier of the task that transfers its
priority.

```
/*****************************************************************
              BPIP semaphore data structure
*****************************************************************/

typedef struct{
    int state;
    int value;
    int holder_id; // new field
    int transfer_id; // new field
    task_queue_t task_queue;
} semaphore_t; semaphore_t *semaphore_ptr_t;

semaphore_t semaphore_tab_t[N_SEM];
```

The new system call to wait on a semaphore is *pip_wait_sema* shown below. This procedure tests whether any task is holding the semaphore and if the holder has a lower priority than the caller, it transfers the priority of the caller to the holder and reschedules to enable the holder to finish executing its critical section. The *transfer_id* in the semaphore is updated to be used in *signal* call to restore priorities. The lower priority task waiting for its turn can now be scheduled by the scheduler since its priority is now raised.

```
/************************************************************
                wait on a semaphore using BPIP
*************************************************************/
int pip_wait_sema(ushort sem_id){

    semaphore_ptr_t sem_pt;
    task_ptr_t holder_pt;
    ushort holder,

    if (sem_id < 0 || sem_id >= system_tab.N_SEM)
        return(ERR_RANGE)
    sem_pt=&(semaphore_tab[sem_id]);
    sem_pt->value--;
    if (sem_pt->value < 0) {
        if (sem_pt->transfer_id==0) {   // new lines start
            holder=sem_pt->holder_id;
            holder_pt=&task_tab[holder];
            if (holder_pt->priority < current_pt->priority) {
                temp=holder_pt->priority;
                holder_pt->priority=current_pt->priority;
                current_pt->priority=temp;
                sem_pt->transfer_id=current_tid;
                Schedule();
        } // new lines end
            insert_task(sem_pt->task_queue, current_pt);
            block(current_tid);
        }
    }
    sem_pt->holder_id=current_tid;
}
```

The signaling of a semaphore in DRTK is done by the *pip_signal_sema* below which increments semaphore count and if there are tasks waiting on the semaphore queue, it restores its original priority by exchanging it with the priority of the task which is kept at the *transfer_id* of the semaphore. We need to *take* the task somewhere from the semaphore queue this time since there may have been other tasks that have performed a *wait* on this semaphore between the *wait* and *signal* operations. If *transfer_id* field is empty, normal *signal* operation is carried.

```
/****************************************************************
                    signal a semaphore
****************************************************************/

int pip_signal_sema(ushort sem_id){

    semaphore_ptr_t sem_pt;
    task_ptr_t task_pt;

    if (sem_id < 0 || sem_id >= system_tab.N_SEM)
        return(ERR_RANGE)
    sem_pt=&(semaphore_tab[sem_id]);
    sem_pt->value++;
    if (sem_pt->value <= 0) {
        if(sem_pt->transfer_id!=0) {      // new lines start
            task_pt=&(task_tab[sem_pt->transfer_id]);
            temp=current_pt->priority;
            current_pt->priority=task_pt->priority;
            task_pt->priority=temp;
            take_task(sem_pt->task_queue, task_pt);
            unblock(task_pt->id, YES);   // new lines end
        }
        else {
            task_pt=dequeue_task(sem_pt->task_queue);
            unblock_task(task_pt->task_id, YES);
        }
    }
}
```

8.5 Review Questions

1. What is the main idea of the LDF algorithm used for scheduling dependent tasks?
2. What is the main difference between the EDF algorithms for independent tasks and dependent tasks?
3. Describe the priority inversion problem.
4. What was the problem with the Mars Pathfinder project?
5. What is the main idea of the BPIP?
6. How may a deadlock be generated in the BPIP?
7. What are the main problems of the BPIP?
8. What is the main idea of the OPCP?
9. How are the main problems of BPIP are addressed in OPCP?

8.6 Chapter Notes

We reviewed the main methods for dependent task scheduling in this chapter. We considered two distinct cases; aperiodic real-time tasks with deadlines and precedence relationships, and aperiodic tasks that share resources. We described algorithms for the first case which can be used for periodic tasks as well when the deadlines are considered as periods of tasks. The input to these algorithms is a DAG showing the precedence among tasks. The LDF algorithm starts from one of the leaf tasks of the DAG and puts this task in a stack. Any task that has no successor with the latest deadline among such tasks is pushed onto the stack until all tasks are processed. The order of execution is then the order of tasks popped from the stack. The MEDF algorithm starts working in the reverse direction by selecting a task that has no predecessors and has the earliest deadline among such tasks to be scheduled first. It continues until all tasks are queued in an FCFS queue and the order of scheduling is the position of tasks in this queue. Both algorithms require $O(n^2)$ time.

We then analyzed the priority inversion problem when real-time tasks share resources. This problem, in fact, occurred in Mars Pathfinder Project in 1985. The main protocols to remedy this issue are the priority inheritance and priority ceiling protocols. The BPIP provides transferring of a high- priority task to a lower priority task that is holding a resource and hence, the resource holding task can finish releasing the resource. This protocol has few drawbacks; a high- priority task may wait for many critical section executions by lower priority tasks and it may cause deadlocks. The priority ceiling protocols can be used to provide deadlock-free operation in which a high-priority task may wait at most one critical section executed by a lower priority task. We showed the implementation of the LDF and MEDF algorithms and the BPIP and the OPCP in DRTK. The BPIP and OPCP protocols required adding new fields to semaphore structure and modifying *wait* and *signal* system calls to semaphores. Comparison of non-preemptive execution, BPIP and OPCP are depicted in Table 8.2 showing number of blockings and deadlock condition.

8.7 Exercises

1. An example aperiodic dependent task set with absolute deadlines is given in Table 8.3. Run the LDF algorithm in this set and check whether all tasks meet their deadlines with this algorithm.

Table 8.2 Protocol comparison

Protocol	Bounded PI	Blocked at most once	Deadlock free
Non-preemptive	$\checkmark$	$\checkmark_a$	$\checkmark_a$
BPIP	$\checkmark$	–	–
OPCP	$\checkmark$	$\checkmark$	$\checkmark$

[a]Only if task does not block in the critical section

Table 8.3 An example task set

τ_i	C_i	d_i	Predecessors
1	2	7	–
2	1	5	1
3	3	8	1
4	1	4	2
5	1	12	2, 4
6	2	10	1, 4

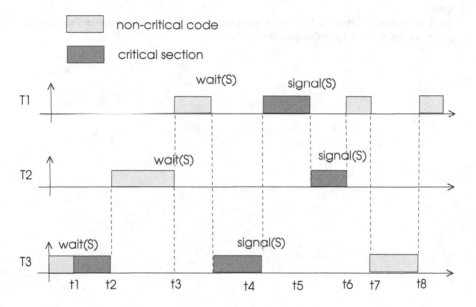

Fig. 8.11 Task executions for Exercise 3

2. Run the MEDF algorithm for the dependent task set of Fig. 8.1 and check whether all tasks meet their deadlines with this algorithm.

3. A task set $\mathcal{T} = \{\tau_1, \tau_2, \tau_3\}$ has the following execution characteristics at increasing times $t_1, t_2, \ldots, t_8$ shown in Fig. 8.11 and use a resource protected by semaphore S_1. Describe the events at each time point and propose a solution using the BPIP. Sketch the Gantt chart of the solution.

4. The non-preemptive protocol to avoid priority inversion may be implemented by assigning the highest available priority to the calling task. Hence, it will be executed without preemption, preventing priority inversion. Show how this protocol can be implemented in DRTK by writing the necessary code modifications.

5. The Immediate Priority Inheritance Protocol works as follows. Any task that wants to enter a critical section performs a wait on the semaphore protecting the resource. If this call is successful, it inherits the highest ceiling of all semaphores

in the system, ensuring it will not be preempted by any other task during its critical section. Implement new *wait* and *signal* system calls in DRTK that exhibit property of this protocol.

References

1. Mars Pathfinder Official Website. http://www.nasa.gov/mission_pages/mars-pathfinder/index. html
2. Sha L, Rajkumar R, Lehoczky JP (1990) Priority inheritance protocols: an approach to real-time synchronization. IEEE Trans Comput 39(9):1175–1185

Multiprocessor and Distributed Real-Time Scheduling

9

9.1 Introduction

We have considered scheduling of tasks in a uniprocessor system up to now and we analyzed few commonly used algorithms for periodic and aperiodic tasks. Scheduling of real-time tasks on multiprocessors and distributed hardware is a more difficult problem than the uniprocessor case. In fact, finding an optimal schedule for given real-time tasks on a multiprocessor or a distributed system is an NP-hard problem and hence, heuristic algorithms are commonly employed to find suboptimal solutions.

When scheduling real-time tasks on multiprocessors, inter-task communication costs may be ignored since communication commonly is accomplished through shared memory, and high-speed parallel *memory buses* are used to transfer data even when the communication involves copying messages from one local memory of a processor to another. The inter-task communication costs are comparable to task execution times in distributed real-time systems since messages are transferred over a communication network and hence, these costs cannot be ignored in such systems. Moreover, the precedence relationship among tasks presents another issue to be solved. Finally, resource sharing adds another level of difficulty when scheduling real-time or non-real-time tasks in a multiprocessor or distributed system.

Scheduling real-time tasks with known properties in multiprocessor or distributed systems is accomplished in two steps; allocation of tasks to processors and scheduling of tasks in the processors and these two steps can be interleaved in some cases.

Our goal in this chapter is to inspect scheduling methods for a multiprocessor system and a distributed real-time system. We will assume interprocess communication costs are negligible in the former due to the tight coupling of processors but we need to consider message delays over the communication network in a distributed system. We will consider independent tasks which do not share resources in all cases unless otherwise specified.

© Springer Nature Switzerland AG 2019
K. Erciyes, *Distributed Real-Time Systems*, Computer Communications
and Networks, https://doi.org/10.1007/978-3-030-22570-4_9

9.2 Multiprocessor Scheduling

The goal of a multiprocessor real-time scheduling algorithm is to assign n tasks of the task set $T = \{\tau_1, \ldots, \tau_n\}$ to m processors of the set $P = \{P_1, \ldots, P_m\}$ such that deadline of each task is met and the load is evenly distributed to the processors. Note that the aim of a general non-real-time multiprocessor scheduling algorithm is mainly to provide a load balance. It is shown in [10] that the multiprocessor scheduling is an NP-Hard problem.

The hardware may consist of *homogeneous* processors in which case the execution time of a task is independent of the processor that it runs, or *heterogenous* processors with tasks having different execution times on processors. For ease of analysis, we will assume the processors are identical. We can have three types of multiprocessor scheduling in general as *partitioned scheduling*, *global scheduling*, and *semi-partitioned scheduling*. Partitioned scheduling is performed offline by assigning tasks to processors after which each processor applies a scheduling algorithm to its assigned tasks as illustrated in Fig. 9.1a. Any uniprocessor scheduling algorithms we have seen in Chap. 7 can be used for local scheduling in partitioned scheduling approach.

Global scheduling takes a different approach by allocating the n highest priority tasks to m processors at any time. It is, therefore, a dynamic approach where any incoming task to the system is scheduled online, based on some scheduling criteria as shown in Fig. 9.1b. There is a single ready queue of tasks in contrast to m ready queues for m processors of the partitioned method.

9.2.1 Partitioned Scheduling

In this mode, we will assume all instances of a task are executed on the same processor as commonly done in practice. We will assume each processor has its private queue of ready tasks and once a task τ_i is assigned to a processor P_j, all instances of τ_i will execute on the processor P_j. Task migrations are not allowed in this approach. Our aim is to partition the task set to m groups such that each group of tasks run on a distinct processor. Scheduling of tasks is done individually by each processor in this case as stated before. We can implement most of the uniprocessor scheduling algorithms and resource management protocols in this method.

9.2.1.1 Task Allocation

The first step in the partitioning scheduling is the assignment of tasks to processors of the multiprocessor system. This step is followed by implementing a uniprocessor scheduling algorithm in each processor. The success criteria is the feasible scheduling of tasks in all processors. The stopping criteria may be defined in several ways; such as failing after k attempts as shown in the following steps of this procedure:

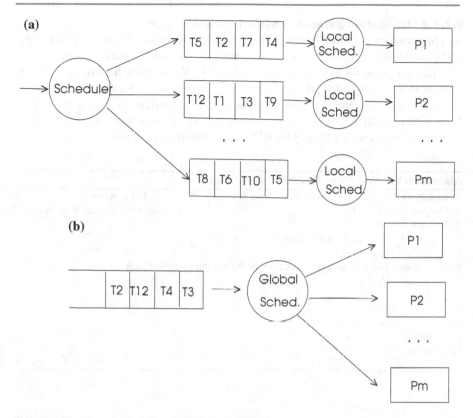

Fig. 9.1 **a** Partitioned scheduling. **b** Global scheduling

1. **Input**: task set $\mathcal{T} = \{\tau_1, \tau_2, \ldots, \tau_n\}$,
2. m processors $\mathcal{P} = \{P_1, \ldots, P_2\}$
3. Allocate tasks to processors with mapping $\mathcal{M} : \mathcal{T} \rightarrow \mathcal{P}$
4. Perform scheduling of tasks at each processor with $\mathcal{T}_i$ is mapped to P_i.
5. **If** schedules on all processors are feasible
6.   **Output** schedule $\mathcal{S} = \{S_1, S_2, \ldots, S_m\}$ where S_i is the schedule of a task set $\mathcal{T}_i \subset \mathcal{T}$ on processor P_i
7. **Else If** stopping criterion is not met goto 3
8. **End If**
9. **Output** failure

We need to check whether the allocation at step 3 provides feasible schedules for all task sets assigned to each processor in the system. If this fails, we may try a different assignment and in the worst-case, the number of processors may be increased. We will describe three sample algorithms for task allocation in partitioned scheduling; utilization balancing algorithm, EDF with first-fit, and RM with first-fit.

9.2.1.2 Utilization Balancing Assignment with EDF

As a first attempt, tasks may be assigned to processors so that utilization of processors are balanced. An unassigned task τ_i is selected from the task set and is assigned to the least utilized processor as shown in Algorithm 9.1. We test the scheduling condition with the task τ_i assigned to processor i; if this fails, we need to have a new processor and hence increment the processor count in line 9. The utilization of a processor i is U_i and the utilization of a task τ_i is u_i. Note that we check whether $U_j + u_i$ is less than unity, hence, we are testing the EDF schedulability condition.

Algorithm 9.1 *Utilization Balancing Algorithm*

1: **Input:** set $\mathcal{T} = \{\tau_1, \ldots, \tau_n\}$ of n tasks, a set $\mathcal{P} = \{P_1, P_2, \ldots, P_m\}$ of processors
2: **Output:** $\mathcal{M} : \mathcal{T} \rightarrow \mathcal{P}$ ▷ assignment of tasks to processors
3: **for all** $\tau_i \in \mathcal{T}$ **do**
4:     **select** P_j such that U_i is minimum
5:     **if** $(U_j + u_i) < 1$ **then**
6:         **assign** task τ_i to the processor P_j with the minimum current load
7:             $U_j \leftarrow U_j + u_i$
8:     **else**
9:             $m \leftarrow m + 1$
10:             $U_m \leftarrow u_i$
11:     **end if**
12: **end for**

9.2.1.3 Bin-Packing Algorithms

Bin-packing method aims to place a number of items of varying sizes to a given number of bins efficiently [6]. This approach can be used for the task allocation step of the partitioning method as we will see. It can be defined formally as follows:

Definition 9.1 (*bin-packing problem*). Given a set of n items $\{1, 2, \ldots, n\}$ with item i having size $s_i \in \{0, 1\}$, find the minimum number of bins with capacity 1 so that all items are placed in the bins.

Solution of this problem is NP-hard [8] and various heuristics can be used to find suboptimal solutions.

- *First-fit*: The bin with the lowest index that can accommodate the item in sequence is selected.
- *Best-fit*: The bin with the smallest capacity that can accommodate the item in sequence is selected.
- *Worst-fit*: The bin with the maximum capacity that can accommodate the item in sequence is selected.

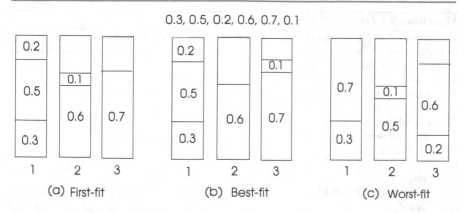

Fig. 9.2 Bin-packing algorithms example

The working of these algorithms is illustrated in Fig. 9.2 with 3 bins of unity capacity each and six items presented in the order 0.3, 0.5, 0.2, 0.6, 0.7, and 0.1.

Bin-packing algorithms may be used for multiprocessor scheduling [12]. In this case, the processors are the bins and the items are the tasks with execution times. Assuming n is the number of tasks and m is the number of processors, Algorithm 9.2 displays the pseudocode for the Next-Fit algorithm to place tasks onto the processors.

Algorithm 9.2 *Next-Fit Bin-Packing Algorithm*

1: **Input:** set $\mathcal{T} = \{\tau_1, \ldots, \tau_n\}$ of n tasks, a set $\mathcal{P} = \{P_1, \ldots, P_m\}$ of processors
2: **Output:** $\mathcal{M} : \mathcal{T} \to P$ ▷ assignment of tasks to processors
3: $i \leftarrow 1; j \leftarrow 1$
4: **while** $i < n \wedge j < m$ **do**
5:     **if** τ_i can be placed on P_j **then**
6:         assign τ_i to processor j
7:         $i \leftarrow i + 1$
8:     **else**
9:         $k \leftarrow k + 1$
10:    **end if**
11: **end while**
12: **if** $i < n$ **then**
13:    **return** INFEASIBLE
14: **end if**

The First-Fit bin-packing algorithm for task assignment has a similar structure but starts from the first processor to fit the current task as shown in Algorithm 9.3 where we assume processor number is constant this time. We search the processors starting from the first one to fit the current task and increment processor count until m is reached.

Algorithm 9.3 *First-Fit Bin-Packing Algorithm*

1: **Input**: set $\mathcal{T} = \{\tau_1, \ldots, \tau_n\}$ of n tasks, a set $P = \{P_1, \ldots, P_m\}$ of processors
2: **Output**: $\mathcal{M} : \mathcal{T} \to P$ ▷ assignment of tasks to processors
3: **for** $i = 1$ to n **do**
4:     $j \leftarrow 1$
5:     **while** $j < m \wedge \tau_i$ cannot be assigned to P_k **do**
6:         $j \leftarrow j + 1$
7:     **end while**
8:     **if** $j < m$ **then**
9:         assign τ_i to P_j
10:        $i \leftarrow i + 1$
11:    **else**
12:        **return** INFEASIBLE
13:    **end if**
14: **end for**

Bin-packing algorithms assume communication costs are negligible due to sharing of data through a global memory or transferring it over high-speed parallel buses.

Let us consider a task set $\mathcal{T} = \{\tau_1, \ldots, \tau_n\}$ to run in a multiprocessor system. First, we need to determine the number of processors needed. This can be determined approximately by considering the total utilization U_T of tasks and checking against the algorithm employed. For example, since we know the EDF algorithm allows a utilization of 1, we need at least $\lceil U_T \rceil$ processors.

9.2.1.4 EDF Algorithm with the First-Fit Bin Packing

The EDF algorithm with the first-fit heuristic may be used to allocate tasks to processors. The EDF algorithm in one processor can provide a utilization of unity and we need to check this property while assigning tasks as shown in Algorithm 9.4. Our aim in this case is to place the tasks on available processors starting from the lowest processor index and checking the utilization of the processor before each assignment. We also deduce the number of processors (j) required.

Theorem 9.1 ([11]). *A task set $\mathcal{T}$ is schedulable by EDF-FF algorithm on m processors if,*

$$U(\mathcal{T}) \leq \frac{m+1}{2} \tag{9.1}$$

If all tasks have a utilization factor C/T smaller than an α value, then worst-case utilization of EDF-FF algorithm is [11]

$$U(m, \beta) = \frac{\beta m + 1}{\beta + 1} \tag{9.2}$$

Algorithm 9.4 *EDF-FF*

1: **Input:** set $T = \{\tau_1, \ldots, \tau_n\}$ of n tasks, a set $P = \{P_1, \ldots, P_m\}$ of processors
2: **Output:** $\mathcal{M} : T \rightarrow P$ ▷ assignment of tasks to processors
3: $i \leftarrow 1; j \leftarrow 1$
4: **while** $i \leq n$ **do**
5:   **if** $U_j + u_i < 1$ **then**   ▷ check EDF criteria
6:    $U_j \leftarrow U_j + u_i$ ▷ assign current task to processor
7:   **else**
8:    $j \leftarrow j + 1$ ▷ increase the number of processors
9:    $U_m \leftarrow u_i$
10:   **end if**
11:   $i \leftarrow i + 1$
12: **end while**
13: **return** (j)

---

where $\beta = \lfloor 1/\alpha \rfloor$. Note that when $\alpha = 1$, then $\beta = 1$ and Eq. 9.2 is reduced to Eq. 9.1.

9.2.1.5 Rate-Monotonic-First-Fit (RM-FF) Algorithm

This algorithm proposed by Dhall and Liu [7] is based on the rate-monotonic approach for the uniprocessor system. It assumes tasks are periodic with deadlines equalling their periods and they are independent. The following theorem specifies the utilization in such a system.

Theorem 9.2 *If a set of m tasks is scheduled according to the rate-monotonic scheduling algorithm, then the minimum achievable utilization factor is* $m(2^{1/m} - 1)$.

The algorithm proposed by the authors consists of the following steps. Tasks are first sorted with respect to their periods and then assigned to processors. Then, tasks assigned to a processor are scheduled using the RM algorithm. For tasks $\tau_1, \ldots, \tau_n$, a task τ_i that is assigned to processor P_j is attempted to be scheduled on P_j and if there is not such a processor, τ_i is assigned to a new processor. Note that this algorithm is similar in structure to Algorithm 9.1 with the test condition adapted to RM algorithm. They showed that RMFF uses about 2.33 U processors in the worst-case where U is the load of the task set under consideration. A possible implementation of this algorithm is shown in Algorithm 9.5.

Analysis

The utilization guarantee bound U_{RMFF} with m processors of RMFF scheduling is provided by Oh and Baker as follows [13]:

$$m(2^{1/2} - 1) \leq U_{RMFF} \tag{9.3}$$

Algorithm 9.5 *RM-FF*

1: **Input**: set $\mathcal{T} = \{\tau_1, \ldots, \tau_n\}$ of n tasks, a set $\mathcal{P} = \{P_1, \ldots, P_m\}$ of processors ▷ m not known
   beforehand
2: **Output: Output**: $F : \mathcal{T} \rightarrow P$ ▷ assignment of tasks to processors
3: **for all** $\tau_i \in \mathcal{T}$ **do**
4:     **select** the lowest previously used j such that P_j can accommodate τ_i based on RM utilization
       test.
5:     **assign** τ_i to P_j
6:     **if** this is not possible **then**
7:         **add** a new processor to the processor set
8:     **end if**
9: **end for**

---

which means a task set with a maximum utilization of about 41% of the total
processor capacity is possible in a multiprocessor platform using this algorithm.

9.2.2 Global Scheduling

This type of scheduling is characterized by a single ready queue of tasks which is used
to assign tasks to nodes as noted before. Also, tasks may be allowed to migrate from
one node to another. We distinguish three cases of migration in general, allowing
job-level migration or *task instance migration* only means an instance of a task may
run on any processor but a started job is not allowed to migrate to another processor.
On the other hand, *task-level migration* allows a task to run on any processor at any
time.

In global scheduling, there is a single global ready queue of tasks and a ready
task may be assigned to a processor with the current least load. Two types of global
scheduling algorithms may be classified; a uniprocessor scheduling algorithm such
as RM or EDF may be adapted or a new algorithm may be designed. We will look
into both approaches. Two common cases of using an existing algorithm may be
distinguished as follows:

- *Global EDF*: The scheduler always selects the m tasks in the queue with the
  shortest deadlines to be scheduled on m processors.
- *Global RM*: The scheduler always selects the m tasks in the queue with the highest
  priorities based on their RM criteria to be scheduled on m processors.

The main problem with this method is the inability to use existing uniprocessor
scheduling algorithms. However, efficient use of processing power may be attained
with appropriate algorithms.

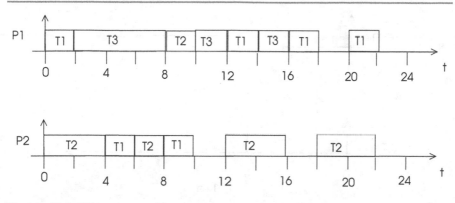

Fig. 9.3 Global RM example

9.2.2.1   Global Rate-Monotonic Algorithm

Consider the task set $\mathcal{T} = \{\tau_1(2, 4), \tau_2(4, 6), \tau_3(8, 24)\}$ that needs to be scheduled to two processors P_1 and P_2. The Gantt chart of a possible schedule is depicted in Fig. 9.3 by assigning priorities to tasks with respect to their periods. At any scheduling point, we select the task with the highest priority to run and this may result in preempting a lower priority task, for example, at time $t = 8$, τ_2 is ready to run and τ_3 is preempted. Note that we allow task migration in this schedule.

9.2.2.2   Anomalies

The following types of anomalies may be present when Global RM or Global EDF algorithms are used in a multiprocessor system.

- Periodic task sets with utilization close to 1 cannot be scheduled with Global RM or Global EDF. This fact is known as Dhall Effect [7]. Consider the case with $n = m + 1$, $\forall \tau_i$, $P_i = 1$ and $C_i = 2\epsilon$, $\forall 1 \le i \le m$. $P_{m+1} = 1 + \epsilon$, $C_{m+1} = 1$, $u_{m+1} = 1/(1 + \epsilon)$. The schedule of these tasks using Global RM is depicted in Fig. 9.4. It can be seen that τ_{m+1} misses its deadline although its utilization is close to unity.
- Increasing period duration of fixed priority tasks may cause missing of deadlines for some tasks [1].

9.2.2.3   The Proportionate Fair Scheduling Algorithm

The proportionate fair (P-Fair) algorithm designed for periodic hard real-time tasks is one of the first optimal real-time multiprocessor scheduling algorithms [3]. The main assumptions in this algorithm are that there are m identical processors, tasks can be preempted and tasks can migrate at no cost. A task is divided into fixed size *subtasks* and the period of a task is divided into possibly overlapping *windows*.

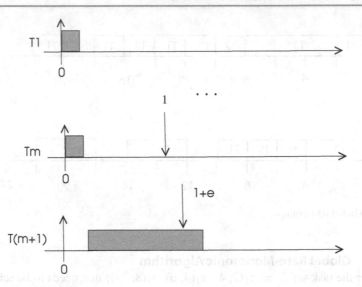

Fig. 9.4 Dhall Effect example

P-fair scheduling algorithms assign priorities to subtasks and attempt to schedule each subtask in its associated window. This way, each job meets its deadline. The P-fair scheduling problem is converted to integral network flow problem and it was shown that $\sum_{i=1}^{n} C_i/P_i \leq m$ where m is the number of identical processors, is the necessary and sufficient condition for a feasible P-fair scheduling [3].

The PF [4] and PD [5] algorithms presented two versions of P-fair algorithm with PD being more efficient. A major disadvantage of P-fair algorithms is the cost related to frequent preemption and migration of subtasks and also, they provide a feasible schedule for periodic tasks. The PD2 algorithm [1] simplifies the priority definition of the P-fair algorithms.

9.3 Distributed Scheduling

Distributed scheduling of tasks in a real-time system commonly has the following goals:

- We need to ensure a priori feasible scheduling of tasks with known characteristics. These tasks are commonly hard real-time periodic tasks with known computation times.
- When an aperiodic task or a sporadic task with a deadline arrives online, we need to perform admission test. Tasks that pass the test are allowed to the system.
- We try to maintain load balancing in the system by migrating tasks from heavily loaded nodes to lightly loaded nodes.

We have seen how partitioning may provide feasible schedules for tasks with known characteristics. In this section, we will review approaches to schedule tasks that arrive asynchronously at runtime by balancing the load in the system at the same time. We first describe two simple methods to balance load in a general system and then briefly review two algorithms to schedule real-time tasks using load balancing.

9.3.1 Load Balancing

Dynamic task scheduling is centered around the concept of *task migration* which is basically transfer of a task with its current environment from a heavily loaded node to another node of the distributed system. Task migration is quite complicated due to the hardness of transferring the environment such as the local and global memory, etc. Another point to consider is whether to transfer a task during its execution, for example, when it is preempted waiting for a resource; or to allow migration only for tasks that have not started execution. We will assume the latter while illustrating two simple dynamic load balancing protocols described in the next sections. We will further assume a critical task τ_i has a number of copies τ_{ij} for $i = 1$ to k, and hence we do not need to transfer the code of a task; we simply need to initiate a task copy at a node with low load.

Yet another point of concern is how to determine the load at a node of a distributed real-time system. A commonly adopted strategy is to determine the number of ready tasks in the ready queues of the scheduler. There can be weights associated with the queues and the tasks themselves. We will assume deadlines of periodic hard real-time tasks are guaranteed by the local real-time scheduling policies and dynamic scheduling involves migration of aperiodic tasks with soft deadlines or sporadic tasks.

9.3.1.1 Central Load Balancing

The central load balancing method is based on a central node we will call the *supervisor* that monitors the loads at each node periodically. It does so by sending a probe message to each node and receiving the local loads during the start of each period. If there exists a low and a high load node pair, it directs the high node to send load and low node to receive load as shown in Algorithm 9.6.

The ordinary node, on the other hand, has a system task which waits forever to receive the polling message *probe* from the supervisor and it sends its current load status to the supervisor when this message is received. Based on its load, it can then wait for load transfer or it can send its load as directed by the supervisor again as illustrated in Algorithm 9.7.

This approach eases the burden on ordinary nodes but as in every protocol that relies on a central entity, two problems faced are that the supervisor may become a bottleneck when the number of nodes in the system is large and second, the supervisor is a single point of failure and in case it fails, we need to elect a new supervisor by the use of a leader election algorithm.

Algorithm 9.6 *Central Load Balancer Supervisor*

1: **Input**: a set $P = \{P_1, \ldots, P_m\}$ of computing elements
2: **Output**: $\mathcal{M} : \mathcal{T} \rightarrow P$ ▷ assignment of tasks to processors 0
3: **while** *true* **do**
4:   upon *timer* expire
5:   **broadcast** *probe* to all computing elements
6:   **receive** *load* from each computing element
7:   **while** there is a *high* node u and *low* node v pair **do**
8:     **send** *load_send(v)* to high node
9:     **send** *load_recv(u)* to low node
10:   **end while**
11:   **set** *timer*
12: **end while**

Algorithm 9.7 *Ordinary Node*

1: **while** *true* **do**
2:   **receive** *probe*
3:   **determine** *my_load*
4:   **send** *my_load* to *supervisor*
5:   **if** *my_load* is high **then**
6:     **wait** to receive *load_send(v)* from *supervisor*
7:     **send** *load* to node v
8:   **else if** *my_load* is low **then**
9:     **wait** to receive *load_send(u)* from *supervisor*
10:     **receive** *load* from node u
11:   **end if**
12: **end while**

9.3.1.2 Distributed Load Balancing

The approach, in this case, is to perform load balancing without the use of a central component due to the disadvantages of employing one. Each node now needs to be aware of the load states of all other nodes. This can be accomplished by each node broadcasting its current load periodically to all other nodes and waiting to receive their loads. In the *receiver-initiated* approach, if a node finds it has low load and discovers a node with high load, it simply asks for transfer from the high node. The high node may receive multiple such requests and it then selects the lowest load node and transfers its load there and rejects all other requesting nodes as shown in Algorithm 9.8.

9.3.2 Focused Addressing and Bidding Method

The focused addressing and bidding (FAB) scheme is an online procedure to schedule tasks in a DRTS [15]. A task set in such a system consists of both critical and noncritical real-time tasks. The initial scheduling of critical tasks guarantees that

---

Algorithm 9.8 *Dynamic Load Balancer Node*

1: **Input**: a set $\mathcal{P} = \{P_1, \ldots, P_m\}$ of computing elements
2: **while** *true* **do**
3:    upon *timer* expire
4:    **broadcast** *my_load* to all computing elements
5:    **receive** *load* from each computing element
6:    **if** *my_load* is *low* and $\exists P_u \in \mathcal{P}$ such that $load(P_u) = low$ **then**
7:       **send** *req_load* to node u
8:       **wait** to receive *load* or *reject* from node u
9:    **else if** *my_load* is *high* **then**
10:      **wait** with *timeout* to receive *req_load* from low nodes
11:      **if** there is at least one *req_load* received **then**
12:         **select** the node v with the lightest load
13:         **send** *load* to node v
14:         **send** *reject* to all other nodes that sent *req_load*
15:      **end if**
16:    **end if**
17: **end while**

---

their deadlines are met so that there is sufficient time reserved for these tasks. The scheduling of noncritical tasks depends on the state of the system. Any such task arriving at a node of the DRTS is first attempted to be scheduled at that node and if this is not possible, a target node that can schedule the arriving task is sought. This scheme consists of *focused addressing* and *bidding* algorithms.

Each node in the system keeps a *status table* showing the list of critical tasks it has been assigned previously by a static scheduling algorithm and other noncritical tasks it may have accepted. Each node has a *load table* containing the surplus capacities of all other nodes in the system. Time is divided into windows of fixed durations and at the end of each window, each node broadcasts the fraction of computer power that it estimates to be free in the next window. Every node receiving this broadcast periodically updates its load table. The load table may not be up to date due to the distributed system.

When a new task τ_i with a deadline arrives at a node P_j, it is attempted to be schedules at that node. This is only possible if surplus power available at the node is greater than the time between the deadline and arrival time of the task. If this is not possible due to node P_j being overloaded, P_j selects a possible node P_s called the *focused processor* and transfers the task τ_i to P_s. However, the surplus information may be out of date as indicated and P_j initiates a procedure called *bidding*, in parallel to communication with the focused node, to increase the possibility that τ_i is scheduled. Bidding is started only if the bidding node estimates there will be In this algorithm, the bidder P_j does the following:

1. Select k nodes with sufficient surplus.
2. Send a request-for-bid (RFB) message that contains the expected execution time, resource requirements and deadline of τ_i to all selected nodes.

3. Send the task to the node that offers the best bid.
4. If bids received are not satisfactory reject the task.

A selected node P_k does the following:

1. Calculate bid which is the likelihood that the task is guaranteed.
2. Send the bid to the bidder if the bid is greater than the minimum required bid.

9.3.3 Buddy Algorithm

In order to decrease the time spent in collecting state information, state probing, and bidding, Buddy algorithm can be employed. The buddy algorithm works similarly to the FAB method by migrating tasks from heavily loaded nodes to lightly loaded ones. The nodes in the system are classified as *Underloaded, Normal-Fully Loaded* and *Overloaded*. The load status of a node is determined by the number of tasks in its scheduler ready queue. A node has a number of nodes associated with it, called its *buddies*. When a node changes its state to *Underloaded* or moves from this state, it informs its buddies of its state. Forming the buddy set needs careful consideration; a large set means many messages and hence significant communication overhead. A small set means finding a receiver of load may not be realized. Choice of thresholds also affect performance; the size of the buddy set, network bandwidth and topology should be considered.

9.3.4 Message Scheduling

In order to achieve end-to-end scheduling of tasks to meet deadlines, message delays over the network need to be considered. Real-time message transfer commonly needs prioritized messages as we reviewed in Chap. 3. As in the case of real-time tasks, real-time communication can be time or event triggered. TTP is an example of the former while CAN is an example of the latter. Time-driven and event-driven systems have different approaches for co-scheduling of real-time tasks and messages. In general, analysis of co-scheduling of tasks and messages is dependent on the communication protocol that is used such as Token Ring or TDMA.

Holistic response time analysis (HRTA) is a schedulability analysis technique to calculate upper bounds on co-scheduling of tasks and messages in a distributed real-time system [9]. Consider the case where a sensor node inputs some data and sends it over the network to the computation node which processes this data and sends some command to an actuator over the network. Holistic response time is the time interval between sensing the data and activating the actuator.

The HRTA method iteratively runs node and network analysis algorithms. The first step is performed by calculating response times of all tasks and messages in the network assuming all jitter values are zero. Then in the second step, the jitter of a message is initialized to the jitter of its sending task computed in the first step and each message receiving task inherits a jitter equal to the response time of the message it receives. The third step is performed by computing the response times of all tasks and messages as before. These computed values are compared against the values obtained in the first step and this procedure continues until the obtained values in steps 1 and 2 are equal. The pseudocode of HRTA algorithm is shown in Algorithm 9.9 where J denotes the jitter value and R denotes the response time.

Algorithm 9.9 *Holistic Response Time Analysis*

1: **Input:** a set $\mathcal{T} = \{\tau_1, \ldots, \tau_n\}$ of tasks
2:       a set $\mathcal{M} = \{m_1, \ldots, m_k\}$ of messages
3: $R \leftarrow 0$ ▷ set total response time to 0
4: **while** *true* **do**
5:   **for all** $\tau_i \in \mathcal{T} \wedge m_i \in \mathcal{M}$ **do**
6:     $J_{m_i} \leftarrow R_{sender_i}$
7:     $J_{receiver_i} \leftarrow R_{m_i}$
8:     **compute** response times of all messages
9:     **compute** response times of all tasks
10:     **if** $R_i! = R_{i-1}$ **then**
11:       $R_{i-1} = R_i$
12:     **else** break;
13:     **end if**
14:   **end for**
15: **end while**

The HRTA was first proposed in [9] for event-driven tasks connected over a TDMA network and later was enhanced by considering dynamic task offsets. In general, messages can be static with known activation time and duration, or dynamic with asynchronous activation. The case of a time-driven system with static messages is considered in [14] to construct a schedule for time-driven tasks and static messages, and also provide schedulability analysis for event-driven tasks and dynamic messages in this system.

9.4 DRTK Implementation

We provide the central load balancing task and the distributed load balancing task for DRTK implementation. We will parse the data area of the transport layer for load balancing tasks to contain the load value and the sender and the receiver of the load values as in the following message type. This message type will be implemented as the union part of the *data* part of the *data_unit_t*.

```
/*****************************************************************
                    Cluster Message Structure
*****************************************************************/

typedef struct {
          ushort load_sender;
          ushort load_receiver;
          ushort load;
          } load_msg_t;
```

9.4.1 Central Load Balancing Tasks

There are two types of tasks for this implementation; the central task and the ordinary
tasks. The central task delays itself and is awaken periodically after which it broadcasts a frame with type LOAD_CHECK to get load state of all nodes in the system.
When all the other nodes send their status, the central task can decide which node
can receive load from which other one and sends the messages for transfer to these
nodes. The *find_load* routine simply counts the number of tasks in a ready queue to
evaluate the load of the node as high or low. The actual task transfer routines are out
of our scope and are not shown. We assume fault-free operation and all nodes are
working correctly.

```
/*****************************************************************
                      Central Load Balancer
*****************************************************************/
/* central_load.c */

#define HIGH_LOAD       120
#define LOW_LOAD         20

#define LOAD_CHECK        1
#define SEND_LOAD         2
#define RECV_LOAD         3
#define LOAD_STATUS       4
#define REQ_LOAD          3
#define LOAD_SENDING      4

TASK Central_Load() {

data_unit_ptr_t received_msg_pts[System_Tab.N_NODES],
data_pt, data_pt1, data_pt2;
ushort load, sender1, sender2, flag=0;

while(TRUE) {
  if(current_tid=System_Tab.Central_Load_id) { // server
    delay_task(current_id, System_Tab.DELAY_TIME);
```

```
      data_pt=get_data_unit(System_Tab.Net_Pool);
      data_pt->MAC_header.sender_id=System_Tab.this_node;
      data_pt->MAC_header.type=BROADCAST;
      data_pt->TL_header.type=LOAD_CHECK;
      data_pt->TL_header.sender_id=System_Tab.Central_Load_id;
      send_mailbox_notwait(System_Tab.DL_Out_mbox,data_pt);
      for (i=0;i<System_Tab.N_NODES-1;i++){
        data_pt=recv_mbox_wait(&task_tab[current_tid].mailbox_id);
        received_msg_pts[i]=data_pt;
      }
  for (i=0;i<System_Tab.N_NODES-1;i++)
   for (j=0;j<System_Tab.N_NODES-1,j++) {
    if(received_msg_pts[i].TL_header.type==LOW_LOAD
    && received_msg_pts[j].TL_header.type==HIGH_LOAD) {
       low=i;
       high=j;
       flag=1;
    }
    else if(received_msg_pts[i].TL_header.type==HIGH_LOAD
    && received_msg_pts[j].TL_header.type==LOW_LOAD) {
       low=j;
       high-i;
       flag-1;
    }
    if (flag==1) {
    data_pt1=received_msg_pts[low];
    data_pt2=received_msg_pts[high];
    sender1=data_pt1->MAC_header.sender_id;
    sender2=data_pt2->MAC_header.sender_id;
    data_pt2->MAC_header.type=UNICAST;
    data_pt2->MAC_header.sender_id=System_Tab.this_node;
    data_pt2->MAC_header.receiver_id=sender2;
   data_pt2->TL_header.receiver_id=data_pt2->TL_header.sender_id;
    data_pt2->TL_header.sender_id=System_Tab.Central_Load_id;
    data_pt2->TL_header.type=SEND_LOAD;
    data_pt2->data.load_msg.receiver_id=sender1;
    send_mailbox_notwait(System_Tab.DL_Out_mbox,data_pt2);
    data_pt1->MAC_header.type=UNICAST;
    data_pt1->MAC_header.sender_id=System_Tab.this_node;
    data_pt1->MAC_header.receiver_id=sender1;
   data_pt1->TL_header.receiver_id=data_pt1->TL_header.sender_id;
    data_pt1->TL_header.sender_id=System_Tab.Dist_Load_id;
    data_pt1->TL_header.type=RECV_LOAD;
    data_pt1->data.load_msg.sender_id=sender2;
    send_mailbox_notwait(System_Tab.DL_Out_mbox,data_pt1);
    }
    flag=0;
  }}
  else { // ordinary nodes
   data_pt=recv_mbox_wait(&task_tab[current_tid].mailbox_id);
   if(data_pt->TL_header.type==LOAD_CHECK) {
```

```
      load=find_load();
      data_pt->MAC_header.type=UNICAST;
      data_pt->MAC_header.sender_id=System_Tab.this_node;
 data_pt->MAC_header.receiver_id=data_pt->MAC_header.sender_id;
   data_pt->TL_header.receiver_id=data_pt->TL_header.sender_id;
   data_pt->TL_header.sender_id=System_Tab.Central_Load_id;
   data_pt->TL_header.type=LOAD_STATUS;
   data_pt->data.load_msg.load=load;
   send_mailbox_notwait(System_Tab.DL_out_mbox,data_pt);
   if(load==LOW_LOAD || load==HIGH_LOAD) {
    data_pt=recv_mailbox_wait(&task_tab[current_tid].mailbox_id);
    if (load==LOW_LOAD && data_pt.TL_header.type==RECV_LOAD)
        // wait to receive load
    data_pt=recv_mailbox_wait(&task_tab[current_tid].mailbox_id);
   else  if (load==HIGH_LOAD && data_pt.TL_header.type==SEND_LOAD)
        // transfer task to low node specified
    }
   }
  }
}
}
```

9.4.2 Distributed Load Balancing Task

In this implementation, we do not have any central nodes and all nodes function equally. Each node is awaken periodically, and then broadcasts its load to all other nodes. In the implemented receiver-initiated approach, a *low* node asks for transfer from a *high* node.

```
/*******************************************************************
                    Distributed Load Balancing
*******************************************************************/
/* distributed_load.c */

data_unit_ptr_t data_pt, received_msg_pts[system_tab.N_NODES],
ushort load;

TASK Dist_Load() {

data_unit_ptr_t data_pt;

while(TRUE){
 delay_task(current_id, System_Tab.DELAY_TIME);
 data_pt=get_data_unit(Sys_Tab.Net_Pool);
 data_pt->MAC_header.type=BROADCAST;
 data_pt->MAC_header.sender_id=System_Tab.this_node;
 data_pt->TL_header.sender_id=System_Tab.Dist_Load_id;
 data_pt->TL_header.type=LOAD_CHECK;
 send_mailbox_notwait(System_Tab.DL_Out_mbox,data_pt);
```

```
for (i=0;i<System_Tab.N_NODES-1;i++){
  data_pt=recv_mbox_wait(&task_tab[current_tid].mailbox_id);
  received_msg_pts[i]=data_pt;
  }
load=find_load();
if(load==LOW_LOAD)
  for (i=0;i<System_Tab.N_NODES-1;i++){
   data_pt=received_msg_pts[i];
   if(data_pt->TL_header.type==HIGH_LOAD){
    data_pt->MAC_header.type=UNICAST;
    data_pt->MAC_header.sender_id=System_Tab.this_node;
   data_pt->MAC_header.receiver_id=data_pt->MAC_header.sender_id;
    data_pt->TL_header.receiver_id=data_pt->TL_header.sender_id;
    data_pt->TL_header.sender_id=System_Tab.Dist_Load_id;
    data_pt->TL_header.type=REQ_LOAD;
    send_mailbox_notwait(System_Tab.DL_Out_mbox,data_pt);
    data_pt=recv_mbox_wait(task_tab[current_tid].mailbox_id);
     if (data_pt->TL_header_type==LOAD_SENDING)
    // receive task and activate it
    break;
   }
   else if(data_pt->TL_header.type==HIGH_LOAD){
 data_pt=recv_mbox_wait_tout(&task_tab[current_tid].mailbox_id);
   if (data_pt->TL_header_type==REQ_LOAD) {
    data_pt->MAC_header.type=UNICAST;
    data_pt->MAC_header.sender_id=System_Tab.this_node;
   data_pt->MAC_header.receiver_id=data_pt->MAC_header.sender_id;
    data_pt->TL_header.receiver_id=data_pt->TL_header.sender_id;
    data_pt->TL_header.sender_id=System_Tab.Dist_Load_id;
    data_pt->TL_header.type=LOAD_SENDING;
    // send task
    break;
   }
  }
 }
}}
```

9.5 Review Questions

1. What are the main methods for multiprocessor scheduling?
2. What are the two phases of partitioned multiprocessor scheduling?
3. Compare partitioned and global multiprocessor scheduling in terms of complexity of algorithms and fair distribution of load.
4. How does a bin-packing algorithm work and what are the main bin-packing algorithms?
5. What is the main idea of the Utilization Balancing Algorithm?
6. How does the EDF with First-Fit work?
7. How does the RM with First-Fit work?

8. What is Dhall Effect?
9. What is the main principle of P-Fair scheduling algorithm?
10. How does Focussed Addressing and Bidding algorithm work?

9.6 Chapter Notes

Two main methods of scheduling tasks in a multiprocessor system are the partitioning and global scheduling schemes. In partitioning method, we need to assign sets of tasks to processors in the first phase and each processor schedules tasks independently from its ready queue in the second phase. On the good side, the basic uniprocessor algorithms such as RM and EDF we have seen can be employed in the second phase; however, the assignment problem is NP-hard necessitating the use of heuristics for suboptimal solutions. Tasks are not allowed to migrate which may result in situations where a feasible schedule could be found if migration was possible.

Global scheduling employs a single ready queue and the function of the scheduler is to ensure the highest priority m tasks run on m processors at any time. Global scheduling with algorithms such as RM and EDF may result in low processor utilizations [7]. New class of algorithms for global scheduling are proposed, the P-fair algorithm with its variations provides an optimal solution to multiprocessor scheduling problem. A survey of real-time multiprocessor scheduling algorithm is presented in [16] and a detailed study of related methods is provided in [2].

Distributed real-time scheduling refers to a real-time system consisting of loosely coupled computational nodes. The network delay is an important parameter to consider in this mode of scheduling. Scheduling algorithms for these systems commonly have two parts; static scheduling of periodic real-time tasks and dynamic scheduling of aperiodic and sporadic tasks. Balancing the load and meeting the deadlines of tasks are the main goals in the latter. Focussed addressing and bidding algorithm provides initial scheduling of hard real-time tasks and a procedure for any incoming tasks that arrive dynamically and buddy algorithm decreases the message traffic in this algorithm. There is room for further research in both partitioning, global scheduling, and hybrid methods. Moreover, we have not considered dependencies and sharing of resources. These add another level of complexity which needs to be addressed.

9.7 Exercises

1. There are 3 bins of unity capacity each and six items in the order 0.2, 0.6, 0.5, 0.1, 0.3, and 0.4. Show how these items can be placed in the bins using First-Fit, Next-Fit, Best-Fit, and Worst-Fit approaches.
2. Modify the Next-Fit code of Algorithm 9.2 assuming we have unlimited number of processors.

3. Given the task set $\tau_1(2, 4)$, $\tau_2(4, 5)$, and $\tau_3(3, 6)$ ($\tau_i(C_i, d_i)$) first check whether this set is schedulable in two processors. If this is possible, work out the EDF-FF scheduling and draw the Gantt chart to show the schedule.

4. Given the task set $\tau_1(6, 12)$, $\tau_2(4, 6)$, and $\tau_3(3, 8)$ ($\tau_i(C_i, T_i)$) work out the RM-FF scheduling assuming all needed processors are available and draw the Gantt chart to show the schedule.

5. Modify Algorithm 9.4 so that EDF with Next-Fit is achieved.

6. Modify Algorithm 9.5 so that RM with Next-Fit is achieved.

References

1. Anderson A, Srinivasan A (2000) Early-release fair scheduling. In: Proceedings of the Euromicro conference on real-time systems, pp 35–43
2. Baruah S, Bertogna M, Buttazzo G (2015) Multiprocessor scheduling for real-time systems (Embedded systems), 2015th edn. Springer embedded systems series
3. Baruah SK et al (1996) Proportionate progress: a notion of fairness in resource allocation. Algorithmica 15(6):600 625
4. Baruah S, Cohen N, Plaxton CG, Varvel D (1996) Proportionate progress: a notion of fairness in resource allocation. Algorithmica 15:600–625
5. Baruah S, Gehrke J, Plaxton CG (1995) Fast scheduling of periodic tasks on multiple resources. In: Proceedings of the international parallel processing symposium, pp 280–288
6. Coffman EG, Galambos G, Martello S, Vigo D (1998) Bin packing approximation algorithms: combinational analysis. In: Du DZ, Pardalos PM (eds). Kluwer Academic Publishers
7. Dhall SK, Liu CL (1978) On a real-time scheduling problem. Oper Res 26(1):127–140
8. Garey M, Johnson D (1979) Computers and intractability. A guide to the theory of NP-completeness. W.H. Freeman & Co., New York
9. Tindell K, Clark J (1994) Holistic schedulability analysis for distributed hard real-time systems. Microprocess Microprogram 40:117–134
10. Leung JYT, Whitehead J (1982) On the complexity of fixed-priority scheduling of periodic real-time tasks. Perform Eval 2:237–250
11. Lopez JM, Diaz JL, Garcia FD (2000) Worst-case utilization bound for EDF scheduling on real-time multiprocessor systems. In: Proceedings of 12th Euromicro conference on real-time systems (EUROMICRO RTS 2000), pp 25–33
12. Morihara I, Ibaraki T, Hasegawa T (1983) Bin packing and multiprocessor scheduling problems with side constraints on job types. Disc Appl Math 6:173–191
13. Oh DI, Baker TP (1998) Utilization bounds for N-processor rate monotone scheduling with static processor assignment. Real Time Syst Int J Time Crit Comput 15:183–192
14. Pop T, Eles P, Peng Z (2003) Schedulability analysis for distributed heterogeneous time/event triggered real-time systems. In: Proceedings of 15th Euromicro conference on real-time systems, pp 257–266
15. Stankovic JA, Ramamritham K, Cheng S (1985) Evaluation of a flexible task scheduling algorithm for distributed hard real-time systems. IEEE Trans Comput C 34(12):1130–1143
16. Zapata OUP, Alvarez PM (2005) EDF and RM multiprocessor scheduling algorithms: survey and performance evaluation. CINVESTAV-IPN, Seccion de Computacion Av, IPN, p 2508

Part IV
Application Design

Software Engineering of Real-Time Systems

10

10.1 Introduction

Software engineering is defined by IEEE as the application of a systematic, disciplined, quantifiable approach to the development, operation, and maintenance of software. Methods for software engineering of a general computer system is well established, however, software engineering for real-time and embedded systems is different than a non-real-time system for a number of reasons. First of all, the hardware platform is to be considered initially as some of the functions required will be performed by the hardware, especially in embedded systems. As a result, the top-down approach used in a general system may not be practical in a real-time system since hardware–software co-design is commonly needed at an initial stage. Second, a real-time system must respond to external events in a specified time frame which means timing analysis is an important component of the design process in such a system and should also be performed at the initial design phase. Some traditional design approaches are still applicable in a real-time system as we will see. Distributed real-time system software engineering brings another level of complexity to the design of these systems as the task communication and synchronization over the network is to be considered as well as assigning tasks to the nodes of the system. In conclusion, there are no well-established methods for software design of real-time systems, whether uniprocessor or distributed, however, some formalism can be achieved by using efficient modeling techniques that incorporate timing analysis.

We start this chapter with the basic and general software engineering concepts of design which can be implemented in a real-time system. We then describe the requirement specification process followed by the timing analysis. Procedural design and object-oriented design are reviewed as applied to real-time systems. The specification and detailed design of real-time systems employ finite-state machines, timed automata, and Petri nets as described.

© Springer Nature Switzerland AG 2019   227
K. Erciyes, *Distributed Real-Time Systems*, Computer Communications
and Networks, https://doi.org/10.1007/978-3-030-22570-4_10

10.2 Software Development Life Cycle

The software development life cycle (SDLC) is the basic framework describing processes that need to be performed at each step, commonly in the form of diagrams. It provides description and order of activities to be performed in a software engineering project. Sequential processes that do not overlap are common in SDLC although overlapping in some processes is inevitable. The following phases described briefly are common to most of the SDLC processes [8]:

- *Requirement Gathering and Analysis*: The goals to be achieved and services to be provided by the system and the restrictions are specified through communications with the user/customer.
- *Design*: The overall system architecture is designed based on the system requirements specification (SRS) document from the previous stage. Design stage typically will produce data design, architecture design, interface design, and procedure design documents.
- *Implementation*: The software to be realized is divided into modules and each team performs the coding of the module it is assigned. Programmers develop the software based on SRS and the design document specifications.
- *Integration and Testing*: Individual software modules are integrated and tested to ensure the whole system works according to the requirements set in the first step.
- *Deployment and Maintenance*: The system is installed and started to operate. During maintenance, the system is modified according to new requirements and any possible errors are corrected.

Some commonly used SDLC models are the Waterfall Model, the Spiral Model and the V-Model as described next.

10.2.1 The Incremental Waterfall Model

The Waterfall Model is a simple and old software engineering high-level design technique that consists of six stages. The flow resembles to waterfall with each stage only completed after the previous one finishes as shown in Fig. 10.1. Note that the feedback from each stage means a previous stage can be modified based on the design of a later stage. The original Waterfall Model does not have these feedbacks in which each phase has to be completed before the next one can start. This simple method can be used from small- to mid-size projects.

The Waterfall Model is easy to understand and implement and is document driven. However, it assumes accurate specification of requirements early in the project which may be unrealistic for many projects. The software is implemented at a relatively later stage in the project causing late discovery of errors.

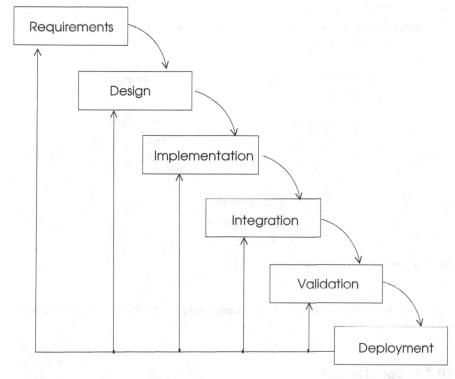

Fig. 10.1 The incremental waterfall model

10.2.2 V-Model

The V-Model provides a graphical representation of software development life cycle in a V-shaped graph containing project development steps. The left part of V contains the requirement analysis, high-level design, and detailed design steps as in the waterfall shown in Fig. 10.2. Bottom of the graph is the implementation where coding is done and the right side has the unit testing, integration testing, system testing, and acceptance testing steps as described below.

- *Unit Testing*: Each module is tested on its own by specific test cases.
- *Integration Testing*: Modules are integrated and tested using the test cases.
- *System Testing*: Complete application is tested with respect to functional and non-functional requirements.
- *User Acceptance Testing*: These tests are done in the user environment to check whether all of the user requirements are satisfied.

The V-Model can be used for software engineering projects with small sizes. Each phase provides specific outcomes and problems with design can be detected at an

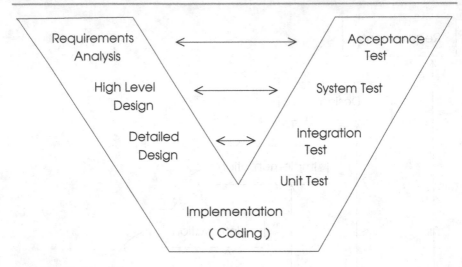

Fig. 10.2 The V-model

early stage. It may not be suitable for large and complex projects where requirements are not clear.

10.2.3 Spiral Model

The Spiral model is significantly more abstract when compared to other software development models. It emphasizes risk analysis and has four phases; problem analysis, risk analysis, realization, and planning. The development goes through these phases in iterations called *spirals*. After circumventing a spiral cycle, these four phases are realized as depicted in Fig. 10.3.

The first phase is the determination of goals and requirements in the cycle followed by the risk analysis where risks are identified and risk reduction techniques are searched. The current product is realized in the third step and planning of the next step is done in the final step. For example, one full cycle may be devoted to the requirement specification in the classical sense followed by high-level design in the second.

This method is commonly used in medium to high-risk projects and when the customer is not certain about the requirements. Also, requirements of the systems may be complex and evaluation may be needed to be precise. Moreover, the user of the system may observe the system behavior at an earlier stage than the Waterfall Model. However, managing phases in this model are complex and it is difficult to determine the termination of the project.

Fig. 10.3 The spiral model

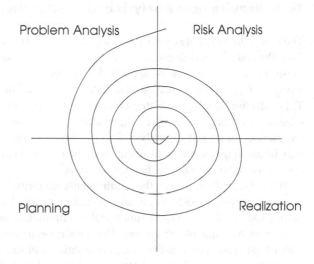

Problem Analysis   Risk Analysis

Planning   Realization

10.3   Software Design of Real-Time Systems

The software design for a real-time and embedded system is different than a general system in various respects. First of all, safety and reliability are of primary concern in a real-time system and these will affect design choices in general. Meeting deadlines is another issue to be handled. Software engineering of real-time and embedded systems should include the following [9]:

- *Choice of Platform*: Selecting the real-time operating system and the hardware has a substantial effect on the performance of the system and hence should be done at very early stages of the design.
- *Input/Response and Timing Analysis*: The inputs and the responses to these inputs must be listed with the required response times. Deadlines for tasks may then be determined. This analysis may be part of the requirement specification.
- *Tasks*: Input from sensors and outputting to actuators, and tasks that run algorithms to do the computation between them are handled by concurrent tasks.
- *Task Scheduling*: Meeting the deadlines of tasks must be ensured.

With these issues in mind, the software development models described as the Waterfall Model, Spiral Model, and the V-Model can be applied to real-time systems.

10.4   Requirement Analysis and Specification

The aim of this stage of software development is to understand the exact requirements from the system through communications with the user and customer and document these. The *user requirements* specify the services to be provided by the system and the *system requirements* are the specifications of the functions to be performed [8]. This software development step consists of two phases; *requirement analysis* and *requirement specification*. The data needed for requirement analysis is obtained from the user/customer and an analysis is done through discussions in the former. There may be ambiguities and contradictions in the requirements of the user/customer and these need to be resolved in the analysis step.

   When the analysis is over, the requirements are formalized in the software requirement specification (SRS) document, sometimes called requirement analysis document (RAD), which contains functional requirements, nonfunctional requirements along with the aims of the system. This task is performed by a *system analyst* who understands and specifies the exact requirements of the customer/system. She will collect data, remove all inconsistencies and anomalies such as contradicting requests and organize and form the SRS document. This document serves as a contract document between the user and the system designer as well as a reference document that guides the high-level design of the system. The acceptance of the product is decided when it conforms to all the requirements in the SRS document. This document specifies *what* needs to be performed by the system leaving *how* to achieve the requirements to design phase. The SRS document should be written with a terminology to be understood by the user, with the specification of the exact requirements. A typical SRS document consists of the following sections:

1. *Introduction*: An overview of the system, its scope, reference to existing systems, the objective of the project, and the success criteria of the project is commonly included in this part.
2. *Functional Requirements*: The high-level functionality of the system is described in this section.
3. *Nonfunctional Requirements*: User requirements such as reliability and performance that are not directly associated with functionality are contained in this part.
4. *Glossary*: It contains a dictionary of terms that are to be understood by the user.

10.5   Timing Analysis

The timing analysis of a real-time system should consider periodic, aperiodic, and sporadic task executions. Aperiodic and sporadic task executions cannot be predicted beforehand, however, a probabilistic analysis for these tasks can be performed. Analysis of timing requirements of periodic tasks should consider the following issues:

- *Task Characteristics*: These include task deadline, execution time, and period. We can have a relatively simpler timing analysis of a real-time system when tasks are independent, periodic and they do not share resources.
- *Task Dependencies*: The execution order of independent tasks must be analyzed. If a task τ_i precedes another task τ_j, τ_i must finish before τ_j can start. The relationship between tasks can be depicted in a task graph and we saw we needed modified task scheduling algorithms when tasks are dependent, in Chap. 8.
- *Resource Sharing*: When tasks share resources, they may get blocked which affects their response times. Also, the priority inversion problem may be encountered which causes a lower priority task to block a higher priority task from execution. We reviewed how priority inheritance protocols may prevent this situation in Chap. 8.

In summary, timing analysis of a real-time system that has dependent, resource-sharing tasks is difficult, however, this analysis is needed to be able to predict the behavior of the system so as to be able to perform acceptance tests for sporadic tasks and find convenient time slots for aperiodic and sporadic tasks without disturbing the meeting of deadlines of periodic tasks.

10.6 Structured Design with Data Flow Diagrams

A data flow diagram (DFD) representation of a software system displays the flow of data through various modules of the system. A DFD consists of the components depicted in Fig. 10.4 where a *process* shown by a circle is a component that performs a system function and *data store* component is a place to keep data for sharing between processes. It is common practice to label the data storage with the plural name given to data coming in and out of the data store. The *external entities* or the *terminators* are shown by rectangles and data flow between various components are depicted by directed arcs. Data flow between the terminators and the system specify the interface of the system to the external world.

The DFD method of system representation is used mainly for its simplicity and many levels of hierarchy it provides. Starting from a very simple diagram that only shows the system as a circle and the external entities, the system circle can be enlarged

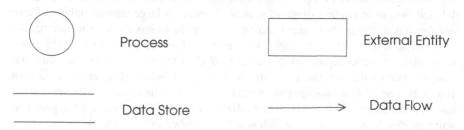

Fig. 10.4 DFD components

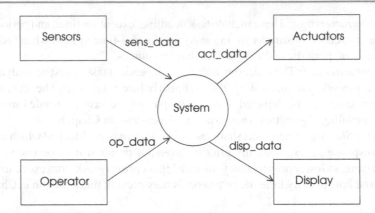

Fig. 10.5 DFD of a typical real-time system

into another DFD which may consist of various processes, data storage, and data flow
between them. The first diagram at level 0 is called the *context diagram* or the *level
0* diagram. The data flow shown by an arc has the label of data that it conveys in a
DFD. The data stores also have names to identify the type of data stored in them. A
data dictionary has the list of all of the labels used in a DFD and also declaration of
any possible composite data elements in terms of the components they contain.

The context diagram of a typical real-time system is depicted in Fig. 10.5. The
external system consists of sensors, actuators, operator interface, and display. Sensors and operator interface provide inputs to the system in terms of data and com-
mands/data respectively and output from the system are directed to actuators and the
display.

Let us consider a simple example where the temperature of an environment is
monitored by this system. Basic functional requirements would then be as follows:

1. Periodically input sensor data, check whether it is in the specified range. If not,
   display alarm and status.
2. If the operator wants to perform an action, activate the required procedure.

A possible implementation of this system using DFD can be realized by refining
the context diagram of Fig. 10.5 to level 1 diagram as shown in Fig. 10.6. The bubble
labeled *System* in context diagram is now shown in a large dashed bubble. There
are four tasks; *Sense_In* to input sensor data, *Op_In* to input operator commands,
Act_Out to output actuator output data, and the *Control* task which is the main
controller. Two data repository places are *all_data* to store all incoming data and
control_data for the *Act_Out* task. Note that the data labeling of the context diagram
is to be preserved. Data storage places can be realized by queues or mailboxes and the
tasks can be modeled by FSMs to be realized by POSIX threads using the practical
software development method which will be described in Sect. 10.10.

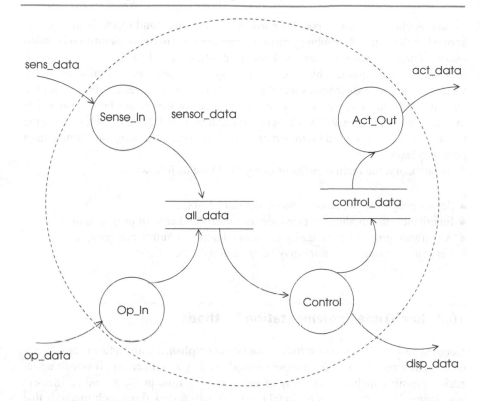

Fig. 10.6 DFD of a typical real-time system

DFDs are practical and simple, however, a drawback with this method is the lack of any control structure. Also, there are no formal techniques to successively form the finer diagrams by decomposing a function to its subfunctions and arrive at the final diagram to start detailed design. This function is carried ad hoc and determined by the experience of the analyst.

10.7 Object-Oriented Design

Object-oriented design (OOD) is developed to eliminate some of the problems encountered in procedural design such as high development costs and inadequate fault management. In this method, the problem is decomposed into entities called *objects* and functions defined on these objects. A *class* defined contains various features such as procedures and data of a logical unit. A software system can then be specified as a structure of collaborating objects. OOD relies on *abstraction* which focuses on what the object does rather than how it is done. For example, *cutting* on object *scissors* and *eating* on object *fork* are analogies to OOD in daily life. Functions called *methods* are specific to objects. A class is a collection of objects of similar types and hence is a general definition of an object and an object is an instance of a class. A class definition specifies attributes of an object and methods that can be applied on the object. Combining attributes and methods in a class definition is called *encapsulation*, which provides restricted access to data and methods of the object. *Inheritance* technique in object oriented design provides hierarchically acquiring variables and methods from a superclass. A *derived class* shares attributes with the class it is derived. A new class may be formed by adding extra attributes to an existing class providing *reusability* which is another favorable feature of OOD. The same function name can be used to implement different operations using different data in *polymorphism*.

In summary, the main benefits of using OOD are as follows:

- Secure programs can be designed using *data hiding*.
- Inheritance and modularity provides *reusability* and ease of programming.
- A software project can be easily partitioned across a number of groups.
- Large and complex software may be managed efficiently.

10.8 Real-Time Implementation Methods

Detailed design of real-time software can be accomplished using a few robust methods. A basic requirement from these methods is they should cover all possible scenarios, possibly include some way of incorporating time in them, and easiness in translating the methodology to actual code. We will review three such methods that are commonly used in detailed design of real-time software; finite-state machines, Petri Nets, and the Unified Modeling Language.

10.8.1 Finite State Machines Re-visited

We have reviewed the *finite-state machine* (FSM) model to specify real-time systems briefly in Chap. 3. An FSM consists of a number of states and transitions between these states represented by a digraph. FSMs can be *nondeterministic* or *deterministic*. A deterministic FSM has exactly one transition for any input and a nondeterministic one may have one or more or no transitions for a given input. We will consider only deterministic FSMs in this text. A deterministic FSM is represented by a 6-tuple; I is the set of finite inputs, O is the set of finite outputs, S is a set of finite states, $S_0 \subset S$ is the initial state, δ is the next state function: $I \times S \rightarrow S$, and λ is the output function.

The digraph called the *finite-state diagram* (FSD) is a visual aid to depict the operation of an FSM. The states shown by circles are mutually exclusive and the system can be in only one of the states at any time. The occurrence of an input or a change of value of some parameters are the common causes of an event which triggers a transition between two states of an FSM. A state change may cause a certain output, which is indicated in the label of the transition arc. For example, a

Fig. 10.7 An example FSM diagram

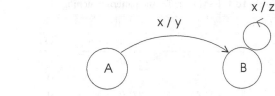

Fig. 10.8 FSM for pattern recognition

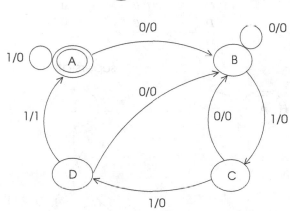

label such as x/y on the arc between states A and B means when x is input, produce output y and change to state B as shown in Fig. 10.7. The same input x may cause a different output z and staying in the same state as depicted.

Two main types of FSMs are the Moore FSM and the Mealy FSM. The output depends only on the state in the former and the output depends on both the input and the state in the latter. In other words, a Mealy FSM produces output only on transitions but not on states. An FSM can be used for *pattern recognition*, which is to detect the occurrence of a predefined pattern on incoming data. An FSM that detects a binary input 0111 and outputs 1 when such input is encountered is shown in Fig. 10.8. There are four states labeled such as A, B, C, and D with A as the initial state. The input set I is $\{0, 1\}$ and the output set O is also $\{0, 1\}$. For each state, we need to define the next state and the output for any input value. Note that failing to receive the third 1 results in returning to state B, not to initial state A, since we already have the first 0 of the pattern.

The finite-state table for this example will have states against all possible inputs and specify the next states and outputs as depicted in Table 10.1.

10.8.1.1   Concurrent Hierarchical Finite-State Machines

A major problem with FSM-based detailed design is the difficulty of specifying and implementing the FSM when the number of states is large. The number of transitions is $O(n^2)$ when n is the number of states, which causes considerable difficulty in visualization and implementation of FSMs. Also, many states of an FSM tend to be similar in practice and a method for reuse of these states would be advantageous.

Table 10.1 FSM table for the pattern matching

Inputs		0		1	
		Next State	Output	Next State	Output
States	A	B	0	A	0
	B	B	0	C	0
	C	B	0	D	0
	D	B	0	A	1

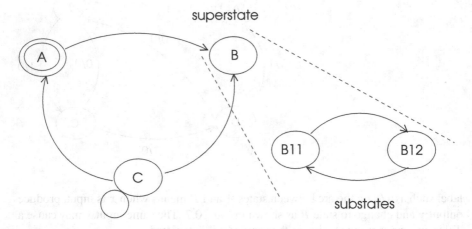

Fig. 10.9 A hierarchical finite-state machine

Statecharts proposed by Harel [4] provides nesting of states to address these problems of FSMs. In this model also known as hierarchical FSMs, a state S_1 of an FSM can be broken down into a number of *substates* $S_{11}, \ldots, S_{1k}$ for example. The state S_1 is called a *superstate* and any event that cannot be handled by a *substate* is delivered to the superstate at the next level of nesting. The substates inherit properties of superstates and define the differences from them, thereby providing reuse of similar states. A hierarchical FSM is depicted in Fig. 10.9 where state B is a superstate and states B_{11} and B_{12} are substates.

A concurrent hierarchical FSM (CHFSM) has a number of FSMs running in parallel each of which may be a hierarchical FSM. The CHFSM model can be used to model a distributed real-time system.

10.8.2 Timed Automata

A *timed automaton* (TA) is a finite-state automaton with added real-time clocks [1], in other words, a TA contains a set of clocks that augment the FSM. The guards in the transition arcs of the FSM diagram now contain clock values which can be

Fig. 10.10 Timed
automaton example

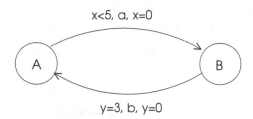

compared against some other values to enable the transition. All clocks increment at the same pace to represent global progress of the TA. The value of a clock can be tested or reset but cannot be assigned a value. A TA with two clocks, x and y is shown in Fig. 10.10. The value of clock should be greater or equal to 5 to enable the transition to state B by performing action a and resetting x. Similarly, returning to state A is enabled when clock $y = 3$ by performing action b and resetting y.

A network of timed automata (NTA) model consists of a set of parallel TAs with a synchronization among them [2]. It is possible to analyze a distributed real-time system using this approach. Each node of such a system is a TA that communicates and synchronizes with other TAs over the network.

10.8.3   Petri Nets

Petri nets provide a mathematical tool to model concurrent and distributed systems [6]. A simple Petri net is a quintuple $PN = (P, T, F, W, M_0)$ where

- $P = \{p_1, p_2, \ldots, p_n\}$ is a finite set of places represented by circles.
- $T = \{t_1, t_2, \ldots, p_m\}$ is a finite set of transitions represented by bars with $P \cup \neq 0$ and $P \cap T = 0$.
- $F \subseteq (P \times T) \cup (T \times P)$ is a finite set of arcs.
- $W : F \rightarrow N^+$ is a weight function.

Token flow among places shows the dynamic behavior of a PN. The components of a PN are depicted in Fig. 10.11 where arcs are directed between a place and a transition, and a transition and a place. An arc has a weight associated with it, with an omitted weight meaning unity. The state of a PN is specified by the distribution of tokens in places.

A PN works according to a firing rule which is stated as follows:

1. *Enabling a transition*: A transition is *enabled* if each input place it has is marked with tokens which have at least the weight of the transition. An enabled transition may or may not *fire*.
2. *Firing a transition*: When a transition is fired, tokens equalling its weight are removed from each of its input places and these are added to its output places. Firing is *atomic*, only one transition fires at a time.

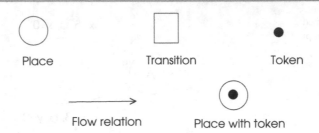

Fig. 10.11 Components of a Petri net

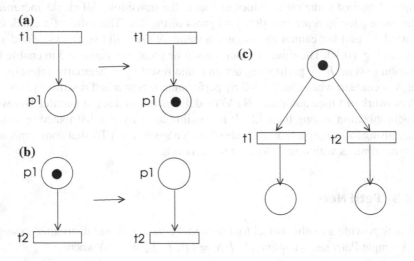

Fig. 10.12 Petri net firing

A transition without any input, called a *source* transition, can fire at any time and produces a token each time it fires as shown in Fig. 10.12a. A transition with no output, called a *sink* may fire when enabled and it consumes the input token when it fires as depicted in Fig. 10.12b. Petri nets are nondeterministic; two transitions with one input having a token compete for the same token and the next transition that will fire is arbitrary. This situation is illustrated in Fig. 10.12c where transition t_1 or t_2 may fire.

Assigning tokens to places of a PN is called *marking*. Modeling of a software system by a PN is achieved by tokens representing resources, places representing states or conditions and transitions representing events or transformations. A PN with three places p_1, p_2, and p_3, and one transition t_1 is shown in Fig. 10.13a. The transition t_1 is enabled since arcs input to it have input places with at least the weights of the arcs. When the transition fires, we have the situation in (b) of the figure where two tokens are placed in the output place p_3 of the transition t_1.

Fig. 10.13 Petri net transitions

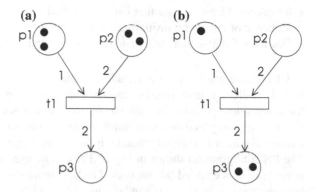

A Petri net with all arcs having unity weights is called a *ordinary* Petri net and we mean an ordinary Petri net when arcs do not have weights. An ordinary PN without any loops and with an initial marking of one token represents a state machine. A PN can be in one of the following states:

- *Current State*: Also called the current marking is the configuration of tokens over the places. For example, the current state of the PN, as shown in Fig. 10.13a for (p_1, p_2, p_3) is (3, 2, 0).

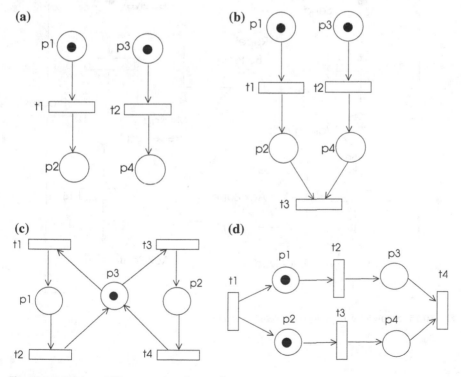

Fig. 10.14 Petri net different modes of executions

- *Reachable State*: A state that can be reached form the current state by firing a sequence of enabled transitions.
- *Deadlock State*: A state where no transition is enabled.

Clearly, we need to ensure a deadlock state is never reached. PNs can be used to model parallel and distributed systems as depicted in Fig. 10.14a where two sequential independent processes are shown. When there is a need for synchronization such that both of two processes must finish for the system to terminate, a transition t_3 is employed which is enabled when both processes finish as depicted in Fig. 10.14b. The PN configuration shown in Fig. 10.14c can be used for mutual exclusion since either t_1 or t_3 is enabled but not both. The last figure shows how parallel processing between transitions t_1 and t_4 can be achieved by a PN.

A communication protocol can be realized using a PN. Figure 10.15 illustrates the implementation of the basic *Stop-and-Wait* protocol by a PN in which the sender sends a message and waits for an acknowledgement before sending the next message. The receiver waits for a message and sends an acknowledgement message to the sender when it receives the message correctly.

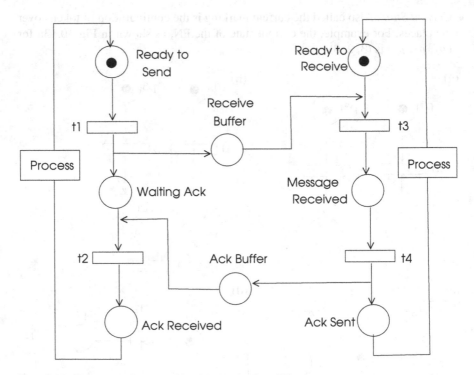

Fig. 10.15 Stop-and-wait protocol implementation by a PN

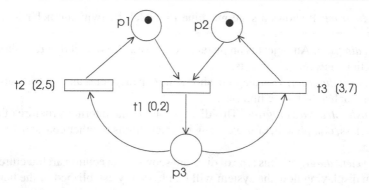

Fig. 10.16 A TPN example

10.8.3.1 Timed Petri Nets

High-level PNs can be formed by adding color, time, and hierarchy to them. A timed (or time) PN (TPN) has t_{min} and t_{max} associated with each transition specifying the earliest and latest times that the transition is enabled [5,10]. An enabled transition can fire if its clock value is in the interval $[t_{min} \ t_{max}]$. A TPN with three places and three transitions with intervals is depicted in Fig. 10.16.

10.9 Real-Time Unified Modeling Language

Unified Modeling Language (UML) is a graphical language that aids analysis, design, development, and implementation of software [3]. It is commonly used in developing OOD of large software systems. Real-time systems require a timely response, preventive measures for faults, provision of safety and reliability as well as quality of service. Real-time UML is basically the implementation of UML with these issues in mind.

10.9.1 UML Diagrams

UML embodies various diagram types to specify software systems. UML diagrams can be classified as *structural* and *behavioral* diagrams. A structural diagram shows what the system must contain with types described below. We will list only the main diagram types of UML as below.

- *Class diagram*: A class diagram shows relationships between the classes used in the system. The operations and attributes of a class may be private, public, or protected in line with the OOD principles. A class diagram is shown as a rectangle with three fields separated by horizontal lines; the name of the class, attributes of the class, and methods of the class.

- *State diagram*: It shows a state machine to model the dynamic behaviour of an object.
- *Object diagram*: An object is an instance of a class. An object diagram shows the interaction between the objects.
- *Component diagram*: This diagram shows the dependencies among software components such as software libraries.
- *Composite structure diagram*: This diagram shows the internal structure of a classifier (class, component, or use case) with interactions to other components of the system.
- *Deployment diagram*: This type of diagram shows the execution architecture of the system displaying how the system will be physically established in the hardware environment.
- *Package diagram*: It shows how components are organized into packages which are logical groupings of software components.
- *Object collaboration diagram*: This type of diagram displays interactions among objects of the system.

The main behavior diagrams display the activity that takes place in the system as shown below.

- *Use-case diagram*: This diagram shows the external view of the system as its connections and interactions to outside world. It helps to display the functional requirements of the system as viewed by the user. It consists of *use cases*, *actors*, and *relationships* like dependency and association between them as shown in Fig. 10.17. An actor represents any entity outside the system such as a person, a group of people or a machine. A possible use-case diagram of a vending machine is depicted in Fig. 10.18.
- *Activity diagram*: This diagram is used to show the dynamic behavior of the system as the procedural flow between objects. The states of an activity diagram are functions. They can be used to model concurrent execution of tasks.
- *State machine diagram*: This type of diagram depicts the control flow in the system as in the FSM model.
- *Object collaboration/communication diagram*: This type of diagram displays interactions among objects of the system.

Fig. 10.17 Use-case diagram components

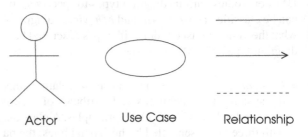

Actor   Use Case   Relationship

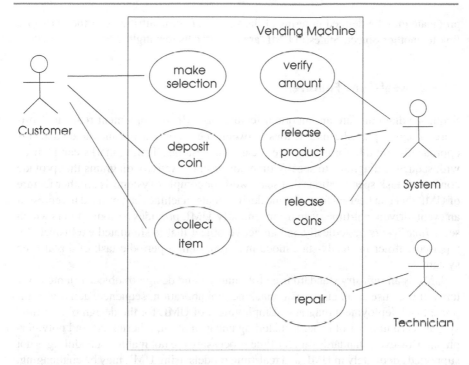

Fig. 10.18 Use-case diagram of a vending machine

- *Sequence diagram*: These diagrams display the interactions between objects with a focus on the ordering of messages. A sequence diagram shows the flow for a specific use case by displaying the calls between different objects. It has a vertical axis showing the message and call sequence in time and the horizontal axis consist of objects that are communicated.

The use of UML entities in the software life cycle can be described as follows:

- *Requirements*: represented by use-case diagrams.
- *Static Structure*: represented by class diagrams.
- *Object Behavior*: represented by a state machine which shows the lifetime of an object.
- *Object Interactions*: represented by activity, sequence, and collaboration diagrams.
- *Physical Implementation*: represented by software modules and their mapping on physical nodes of the system.

A state in UML represents a duration in the life of an object during which the object satisfies some condition, performs some action or waits for an event. A state is commonly associated with a predicate which is true while the state is active. A predicate may be defined in terms of the values of some attributes of the class or a link to another object. States in UML are depicted as rectangles.

10.9.2   Real-Time Features

Sequence diagrams are an important feature of UML to implement real-time software. They display the interactions between the objects using messages and this approach is suitable for event-driven real-time systems. Timing marks can be used with sequence diagrams to specify time constraints. Activity diagrams that provide concurrent task specifications that scale well for complex systems is another feature of UML that can be used for real-time design. State machines can be used to represent an event- driven real-time system conveniently. UML provides periodic timers which send *TimeEvents* periodically and an active object that is simulated exclusively by a periodic timer in steady-state mode may represent a periodic task of a real-time system.

UML is an industry standard used for analysis and design of object-oriented systems. It uses use case, class, state machine, collaboration, sequence, activity, component and deployment diagrams. Employment of UML for the design of real-time system has a number of benefits including encapsulation, inheritance, and polymorphism. However, standards for real-time processing are not mature, scheduling is not supported adequately in UML and real-time models using UML may be challenging.

10.10   A Practical Design and Implementation Method

A practical and novel design and implementation method for small to medium real-time systems can be constructed as follows:

1. *Requirements*: List the requirements from the system as visioned by the user and write the system requirement specification document. This document should specify all the deadlines of hard real-time tasks to be met.
2. *Operating System*: Select a real-time operating system (RTOS) that is suitable for the application. An RTOS with POSIX threads facilities will be useful.
3. *High-Level Design* 1: Design the system using data flow diagram (DFD) method starting with the context diagram at level 0. This diagram should show all external entities and the data flow between these entities and the system.
4. *High-Level Design* 2: Recursively detail the context diagram to level 1 DFD and then to level 2 and so on, preserving data labeling. Continue until a bubble roughly corresponds to a function to be realized with input and output data flows.
5. *Timing Analysis*: Perform timing analysis and if the system to be designed is a DRTS, do distributed scheduling so that deadlines of tasks are met.
6. *Detailed Design*: Design an FSM for each function bubble with states and transitions between the states.

Fig. 10.19 Main steps of the practical design method

7. *Coding* 1: Implement each FSM as a POSIX thread in C language.
8. *Coding* 2: Use any existing POSIX library functions for thread synchronization and communication between the FSMs.
9. *Coding* 3: Use facilities of the RTOS for other functions such as memory management and interrupt handling.

This procedure has been implemented in various real-life applications with success. We will see a case study using this approach and DRTK as the RTOS in Chap. 13. The coarse steps of this method is depicted in Fig. 10.19.

10.11   Review Questions

1. What are the differences between software engineering of a general system and a real-time system?
2. What is the problem with the classical Waterfall Model and how can this be remedied?
3. What is the main characteristic of the V-Model?
4. What is the outcome of the requirement analysis of software engineering?
5. What are the main types of software engineering projects where Spiral Model can be used?
6. What are the main things to consider in the timing analysis of a real-time system?
7. What is the main working principle of DFDs?
8. What are the main benefits of using object-oriented paradigm in the design and implementation of real-time systems?
9. What are the main problems with FSMs and how are these problems solved in CHFSMs?
10. What is the difference between an FSM and Timed Automata and what is a Networked Timed Automata?
11. What are places and transitions in a Petri net and what is the function of a token?
12. What are the main UML diagrams?
13. What is the function of a UML use-case diagram?

10.12   Chapter Notes

We have described basic software engineering concepts as related to real-time systems. Although software engineering methods are well established, there are not any traditional formal software engineering methods of real-time systems whether uniprocessor or distributed. These systems are embedded in many cases, necessitating the co-design of hardware and software at an early stage of the design process. Also, bottom-up approach in the design is commonly pursued in many projects.

The main steps of software engineering are requirements analysis, design, implementation, testing, and maintenance. Various models are used to represent software development life cycle with the main ones as Waterfall Model, V-Model and the Spiral Model. The Waterfall Model is the classical model in which the outcome from a phase finished is used as an input to the next phase. This approach is problematic in practice since realizing a stage may require returning to and modifying a previous stage and hence, the incremental Waterfall Model is introduced, allowing feedback between the development phases. The V-Model presents the process of design in the left side and processes of testing on the right side of a V- shaped graph with implementation at the bottom. The Spiral Model consists of cycles each of which goes through the phases of analysis, risk analysis, realization, and planning. This method can be used in complex systems in which requirements may not be clear in the beginning.

Requirement analysis is typically the first stage of software development. The customer needs are formed into the system requirement specification (SRS) document which forms the basis for high-level design and is also the contract between the designer and the customer/user. High-level design methods may be classified as procedural and object oriented. Data flow diagrams, structured flow charts are the commonly used tools for high-level procedural design and the object-oriented Unified Modeling Language can be used for high-level and detailed design in software engineering projects.

Detailed design methods in real-time systems commonly employ FSMs, Timed Automata, and Timed Petri Nets. An FSM consists of states and transitions between these states. A Timed Automata is basically an FSM augmented with timed transitions. A Petri net consists of places and transitions and a Time Petri Net provides timed transitions between the places. All of these methods may be used to specify and perform detailed design of real-time systems. A distributed real-time system brings another level of complexity to the design; timing analysis should be handled in terms of end-to-end meeting of deadlines and analysis of network delays should also be considered. UML is an object-oriented platform for designing, analysis and implementing software systems. It can be used for real-time systems bringing a number of advantages such as data encapsulation, inheritance, and polymorphism, however, design of predictable systems, scheduling and real-time modeling features of UML are not adequate and mature for this purpose yet.

Lastly, we showed a tested, simple, and practical method to design a real-time system. The requirements specification phase is performed as in a non-real-time system, high-level design is performed by DFDs and level diagrams are recursively formed until a bubble corresponds to a function. This function is then realized by an FSM (or Timed Automata) which can be coded by a POSIX thread. This way, all library functions pertaining to POSIX threads can be accessed and furthermore, an RTOS employed can provide any other needed function such as interrupt handling and clock management. Software engineering is a well-established discipline for non-real-time systems and detailed analysis of software engineering concepts can be found in [9] and [7]. Software engineering methods for real-time systems are still at an early stage of development needing formal and applicable techniques.

10.13   Programming Projects

1. The software for an elevator that goes between two floors is to be realized. Draw the DFD starting from the context diagram. Identify tasks and implement them using POSIX threads.
2. The four types of nucleotides in human DNA are Adenine (A), Thymine (T), Guanine (G), and Cytosine (C). The human genome is to be searched for a nucleotide pattern ACCGTA. Draw the diagram of the FSM and the state transition table to detect this pattern and write the code in C language to implement the FSM.
3. Compare FSM, Timed Automata, and Petri Net modeling of distributed real-time systems in terms of representing time and implementing distributed processing.
4. Draw the use-case diagram of UML for point of sale (POS) terminal with the customer and clerk and the POS machine.

References

1. Alur R, David L, Dill DL (1994) A theory of timed automata. Theor Comput Sci 126:183–235
2. Balaguer S (2012) Concurrency in real-time distributed systems. PhD thesis, Laboratoire Specification et Verification
3. Booch G, Rambaugh JE, Jachobson I (1998) UML user guide. Addison Wesley
4. Harel D (1987) Statecharts: a visual formalism for complex systems. Sci Comput Program 8:231–274
5. Merlin PM (1974) A study of the recoverability of computing systems. PhD thesis, Department of Information and Computer Science, University of California, Irvine, CA
6. Petri CA (1962) Kommunikation mit Automaten. PhD thesis, University of Bonn
7. Pressman RS (2014) Software engineering: a practitioner's approach, 8th edn. McGraw-Hill Education
8. Sommervilee I (2011) Software engineering, 9th edn. Addison Wesley
9. Sommervilee I (2011) Software engineering, 9th edn. Addison Wesley (Chap. 20)
10. Zuberek WM (1991) Timed Petri Nets—definitions, properties, and applications. Microelectron Reliab 31(4):627–644

Real-Time Programming Languages

11

11.1 Introduction

Once the high-level and detailed designs are done, the next stage of development of a real-time system is the coding which is commonly done using a real-time programming language. Such a language is characterized by its facilities for time management, task synchronization and communication, and exception handling and scheduling. Yet, another important property to search is whether the language allows distributed processing. At first glance, it can be seen that these facilities are provided at real-time operating system level as we have reviewed in Chap. 4. However, extension of these capabilities to programming level is practical since some parts of real-time software design are left to the programmer, at the expense of provision of experienced real-time programmers. Furthermore, the code will be portable across different operating systems and will be easier to maintain when a real-time language is used. On the other hand, the model adopted by the operating system may be radically different from the model used in the language making the implementation difficult [2]. Assembly language is closest to the hardware, and many times the real-time programmer needs to write some assembly language patches to access the bare hardware.

There are only few high-level real-time programming languages, and our selected set for review consists of C/Real-time POSIX, Ada, and Java. We start this chapter by the C/Real-time POSIX brief description followed by the reviews of Ada and Java. Our focus in the analysis of these languages is confined to the properties described above that make them suitable for a real-time application.

© Springer Nature Switzerland AG 2019
K. Erciyes, *Distributed Real-Time Systems*, Computer Communications
and Networks, https://doi.org/10.1007/978-3-030-22570-4_11

11.2   Requirements

The requirements from a real-time programming language require additional facilities to a non-real-time language as noted. The common requirements for real-time programming along with non-real-time languages may be listed as follows.

- *Module Management*: Independent development of software is needed in large projects. For this reason, convenient decomposition of the software into modules is a basic requirement from any real-time or non-real-time programming language.
- *Data Encapsulation*: Data needs to be protected against erroneous use. Object-oriented paradigm provides data encapsulation conveniently.

The following properties are more prominent in a real-time language than a non-real-time one.

- *Concurrency Support*: Typically, a real-time system is connected to an external world of asynchronous events. These events require parallel processing which may be realized conveniently by an operating system or a programming language that allows multitasking. Also, task timing and relationship analysis may be done in a multitasking system to perform schedulability analysis. Inter-task synchronization and communication, and sharing of resources are the main issues to be handled in a concurrent system as we reviewed in Chap. 4. The programming languages Ada and Java provide concurrency support, whereas the C programming language with POSIX interface can be used for concurrent applications.
- *Input/Output Support*: The programming language should provide mechanisms to access the I/O hardware using registers and machine-level manipulations.
- *Time Management*: Time is the most precious resource in a real-time system, and a real-time programming language should have facilities for managing timers, and coding of periodic and aperiodic tasks conveniently.
- *Scheduling Support*: Real-time systems employ priority-based scheduling in most cases; hence, the real-time programming language should have facilities to assign priorities to tasks as well as some support for priority-based scheduling.
- *Support for Exception Handling*: Handling exceptions is needed in any software system whether real-time or not. However, the consequences of a fault may be more severe in a real-time system than a non-real-time system; hence, a desirable feature of a real-time programming language is the efficient handling of exceptions.

Note that most of these properties are considered essential in a real-time operating system. Based on the properties outlined above, we will have a short review of the programming languages C/Real-time POSIX, Ada, and Java.

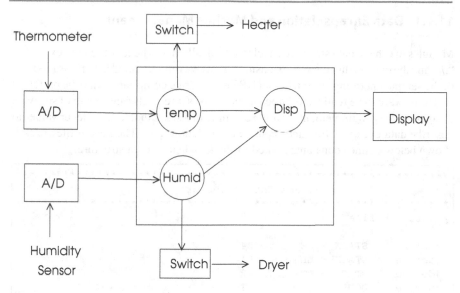

Fig. 11.1 A real-time process control system

11.3   A Real-Time Application

We will describe a real-time application that will be used to show how a programming language may implement the desired functions. The application is intended to monitor heating and humidity in a process control system. There are two sensors to detect the temperature and humidity of the environment. Both inputs go through an A/D converter to the input interface of a real-time computer which has three tasks; the *Temp* for the heat control, *Humid* for the humidity control, and *Disp* for monitoring display as shown in Fig. 11.1. The switches are used to turn the heater or the dryer on or off.

11.4   C/Real-Time POSIX

C is a general-purpose programming language that is commonly used for low-level programming, writing operating system code, and embedded system applications. It has a relatively small number of keywords compared with other programming languages. We will go through the C/POSIX properties associated with the needed attributes of a real-time programming language.

11.4.1 Data Encapsulation and Module Management

Modules in C have data structures declared typically in the header file with extension
"*.h*" and the code in the file with extension "*.c*" as we implemented in various modules
of the sample operating system kernel DRTK. Separate compilation may be preferred
when software to be realized is large and has to be shared between a team of software
designers. We will implement a simple example of a stack that holds integer values to
describe data encapsulation and separate compilation in C. The header file *stack.h*
shown below defines constants related to stack and stack data structure.

```
/*****************************************************************
                    stack data structure
*****************************************************************/
// stack.h file

 #define     STACK_SIZE      1024
 #define     STACK_FULL      -1
 #define     STACK_EMPTY     -2
 #define     DONE            1

 typedef struct {
    int state;
    int data[STACK_SIZE];
    ushort index;
 }stack_t;

 typedef stack_t *stack_ptr_t;
```

Two operations on the stack data type are *push* and *pop* functions which are in
the *stack.c* file which includes the stack header file *stack.h*.

```
/*****************************************************************
                    stack functions
*****************************************************************/
// stack.c file
#include <stack.h>

int push(stack_ptr_t stack_pt, int item){
    if (stack_pt->index == 0)
        return(STACK_FULL);
    stack_pt->data[stack_pt->index]=item;
    stack_pt->index--;
    return(DONE);
}

int pop(stack_ptr_t stack_pt){
    if (stack_pt->index == STACK_SIZE)
        return(STACK_EMPTY);
    stack_pt->index++;
    return(stack_pt->data[stack_pt->index]);
}
```

We have a test program in the file *stack_test.c* which pushes 10 integers from 1 to 10 onto the stack and then pops these one by one and prints each data item.

```
/*****************************************************************
                     main program
*****************************************************************/
// stack_test.c file
 #include <stdio.h>
 #include "stack.h"
void main(){
    int i, data;
    stack_t stack_ex={0,{0},STACK_SIZE-1};
    stack_ptr_t spt=&stack_ex;
    for(i=1; i<=10; i++)
       if(push(spt,i) <0)
          exit(0);
    for(i=1; i<=10; i++)
       { if((data=pop(spt)) <0)
          exit(0);
         printf(" data retrieved: \%d \\n",data);
       }
}
```

We can compile the C source files separately to test whether they contain errors. Final step is to link the compiled files into a single executable file *stack_test* as shown below.

```
gcc -c stack.c
gcc -c stack_test.c
gcc -o stack_test stack_test.o stack.o
```

11.4.2   POSIX Thread Management

We have implemented few examples using POSIX threads; hence, we will only list the main thread management functions of POSIX interface.

```
int pthread_attr_init(pthread_attr_t *attr);

int pthread_create(pthread_t *thread,const pthread_attr_t *att,
void *(*start_routine)(void *), void *arg);

int pthread_join(pthread_t thread, void **value_ptr);

int pthread_exit(void *value_ptr);

pthread_t pthread_self(void);
```

The *pthread_create* function is typically called from the main thread possibly with some input parameters. The main thread waits for a thread it has created by the *pthread_join* call. The *pthread_exit* function is used when a thread finishes execution, and it can also pass the address of a parameter to the thread waiting to join it using this function. The last call is used when a thread needs to learn its identifier. Note that we can pass an integer to a thread during its creation that can be designated as its identifier, different from the identifier that the operating system assigns to the thread. The user-assigned thread identifier is useful when parallel processing by a number of threads is performed and each thread works on a specific part of data based on its identifier. The following code segment illustrates how to assign identifiers and joining of threads. The main thread creates 10 threads and passes the index of the loop as its identifier to the created thread. The outputs of thread computations are gathered in the array *results*.

```
#include <pthread.h>
#define N_THREADS    10
pthread_t tid[N_THREADS];
int results[N_THREADS];

void Thread((void *)me)
{
   // do some computation
   // compute result
   pthread_exit(&result);
}

void main() {
 int i;
 for ( i=1; i<=N_THREADS; i++)
    pthrad_create(&tid[i], NULL, T, (void *)i);
 for ( i=1; i<=N_THREADS; i++)
    pthread_join(&results[i]);
}
```

11.4.2.1   Time Management
A standard way of managing time in C is by using the data *timespec* specified as follows.

```
struct timespec {
time_t tv_sec; // number of seconds
long tv_nsec; // number of nanoseconds
}
```

Setting and getting the time is done by the following functions in POSIX interface.

```
int clock_gettime(clockid_t clock_id, struct timespec *tp);
int clock_settime(clockid_t clock_id, const struct timespec *tp);
int clock_getres(clockid_t clock_id, struct timespe *res);
```

where *clock_t* represents the type of real-time clock to be used. Delaying of a task in RT-POSIX is achieved by the *sleep* or *nanosleep* system calls.

```
unsigned sleep(unsigned seconds);
int nanosleep(const struct timespec *ts, struct timespec *tw)
```

where *ts* specifies the interval thread is suspended. When a sleeping thread is interrupted, its remaining time is written to the structure pointed by *tw* unless *tw* is NULL. A periodic thread can be implemented by the use of *nanosleep* system call as follows where the period is specified as 80+computation time in milliseconds.

```
void *thread(void *arg) {
   struct timespec period;
   period.tv_sec = 0;
   period.tv_nsec = 80 * 1000000; // 80 msec
   while(1) {
   // do some computation
   nanosleep(&period, 0);
   }
}
```

11.4.2.2   Thread Synchronization and Communication

The C/POSIX interface provides two methods of managing concurrency: using UNIX process management with inter-process communication and synchronization methods and by using POSIX threads. We have reviewed both of these concepts in detail in Chap. 4. In general, there are four methods for thread synchronization in POSIX interface.

- Signals
- Mutual exclusion
- Conditional synchronization
- Semaphores

We will briefly review these methods with emphasis on their real-time properties in the next sections.

11.4.2.3 Signals

UNIX processes use signals to synchronize. A process that wants to catch a signal uses *signal* call as follows:

$$\text{signal(int signum, sighandler\_t handler);}$$

where *signum* is the number of signal to be caught and *handler* is the address of the function to be invoked when signal is received. The handler field may be set to SIG_IGN to ignore the signal or to SIG_DFL for default action. Sending a signal to the specified process is done by the *kill* function as below:

$$\text{int kill(pid\_t pid, int signum);}$$

where *pid* is the process identifier of the receiver and *signum* is the signal number. The following call may be used by a thread to send a signal to another thread of the same process.

```
int pthread_kill(pthread_t thread, int signum);
```

The *pause()* system call causes the calling process or thread to wait for a signal. The calls *sigwaitinfo*, *sigtimedwait*, and *sigwait* may be used to wait until one of the signals in the specified set is received. The following example shows how to catch SIGALRM signal generated when time runs out.

```
#include <signal.h>

void my_handler(int sig) {
      signal(SIGALRM, my_handler);
      .... // do interrupt serving
}

main(void) {
  signal(SIGALRM, my_handler);
  while(true) {
    alarm(20);
    ....       // do some work
  }
```

11.4.2.4 Mutual Exclusion

Threads may have critical sections to manipulate globally accessible data and access to these critical sections must be mutually exclusive. POSIX library provides the data structure *pthread_mutex_t* which can be set upon entry to a critical section and reset upon exit. A mutex variable m has to be declared first as below.

$$\text{pthread\_mutex\_t m;}$$

The mutex data structure and the functions associated with mutexes are as follows.

```
int pthread_mutex_init(pthread_mutex_t *mutex, NULL);
int pthread_mutex_lock(pthread_mutex_t *mutex);
int pthread_mutex_trylock(pthread_mutex_t *mutex);
int pthread_mutex_timedlock(pthread_mutex_t *mutex,
    const struct timespec *abstime);
int pthread_mutex_unlock(pthread_mutex_t *mutex);
```

The first function initializes the mutex created, and the other two functions *pthread_mutex_lock* and *pthread_mutex_unlock* are used to lock the mutex and to unlock the mutex, respectively. Note that calls to these two functions lock and unlock a data structure but not other threads. The basic rule is that whenever two or more threads access the same global data, their access should be encapsulated by locking and unlocking a mutex variable to prevent a race condition. The *pthread_mutex_trylock* call attempts to lock a mutex and gives an error if mutex is already locked. The *pthread_mutex_timedlock* call also attempts to lock a mutex and gives an error if mutex cannot be locked in the specified time interval. These two routines may be used by real-time tasks to prevent unpredictable blocking.

11.4.2.5   Conditional Synchronization

A condition variable of type *pthread_cond_t* is used for inter-thread synchronization. A thread may wait on a condition by the call *pthread_cond_wait* to be awakened by another thread by the call *pthread_cond_signal* that signals it. The basic condition functions are as follows.

```
int pthread_cond_init(pthread_mutex_t *mutex, NULL);
int pthread_cond_wait(pthread_mutex_t *mutex);
int pthread_cond_timed_wait(pthread_mutex_t *mutex);
int pthread_cond_signal(pthread_mutex_t *mutex);
int pthread_cond_destroy(pthread_mutex_t *mutex);
```

The timed waiting on a condition is convenient when a real-time thread can wait for a specific condition only for a limited time. The call *pthread_cond_destroy* is used to delete a condition from the system. Note that conditional synchronization is performed on mutex variables.

11.4.2.6   Semaphores

A semaphore in POSIX interface is a data structure that can be used for mutual exclusion and inter-thread and inter-task synchronization. We have reviewed POSIX semaphore implementations in Chap. 4, and hence we will only list the basic semaphore operations here.

```
int sem_init(sem_t *sem, int pshared, unsigned int value);
int sem_wait(sem_t *sem);
int sem_post(sem_t *sem);
int sem_getvalue(sem_t *sem, int *valpt);
int sem_destroy(sem_t *sem);
```

A semaphore is initialized by the *sem_init* system call with an initial *value*, and the variable *pshared* indicates whether it will be shared among other processes that are active. The *sem_wait* system call blocks the caller if the resource is not available and *sem_post* call increments the value of the semaphore and unblocks a waiting thread on the semaphore queue. The last routine simply deletes a semaphore from the system.

11.4.3 Exception Handling and Low-Level Programming

There is no explicit exception handling in the C/POSIX interface; however, a simple way to achieve fault tolerance can be done by the employment of exception handling procedures when a call from a function returns error. Assembly language programming interface in C is provided by the *pragma asm* utility as shown in the code segment below. The assembly instructions encapsulated in the block are executed after which C instructions start to run.

```
extern void test();

void main( void ) {
 ....
pragma asm
 JMP address
pragma endasm
}
```

11.4.4 C/Real-Time POSIX Process Control Implementation

We will now implement the process control of Fig. 11.1 using C/POSIX. The three tasks will be represented by POSIX threads; the *Temp* and *Humid* tasks are time-triggered periodic tasks and the *Disp* task is event-driven. The *Temp* thread is shown below; it is a periodic task awaken at every 200 ms to input and check the temperature data against a low limit value. If the plant temperature is lower than this limit, *Temp* turns the heater switch on.

```
/****************************************************************
                      Temp Task
 ****************************************************************/
#define LOW_T_LIMIT     12
#define INTERVAL_T      0.2

typedef struct {
      int type;
      int alarm_cond;
      float value;
      }data_t;
```

```
typedef struct{
    data_t temp_data;
    data_t humid_data;
    }disp_data_t

disp_data_t disp_data;

void Temp(void *) {

    while(true) {
      sleep(INTERVAL);
      temp_value=Read_T();

      sem_wait(temp_sem);
      if (temp_value<LOW_T_LIMIT) {
         Turn_H_Switch();
         disp_data.temp_data.alarm_cond=ON;
      }
      disp_data.temp_data.value=temp_value;
      sem_post(disp_sem);
    }
}
```

The humidity task *Humid* works similar to the *Temp* task as in the code below, with a different low limit and a period.

```
/*****************************************************************
                     Humid Task
******************************************************************/
#define LOW_LIMIT    12
#define INTERVAL_H   0.1

void *Humid(void *) {

    while(true) {
      sleep(INTERVAL);
      humid_value=Read_H();
      sem_wait(temp_sem);
      if (humid_value<LOW_H_LIMIT) {
         Turn_D_Switch();
         disp_data.humid_data.alarm_cond=ON;
      }
      disp_data.humid_data.value=humid_value;
      sem_post(disp_sem);
    }
}
```

The *Disp* task always waits on its semaphore and when activated, first finds the source of data, then checks the alarm condition, and finally displays data as in the code below.

```
/****************************************************************
                        Disp Task
 ****************************************************************/
#define LOW_LIMIT    12
#define INTERVAL_H   0.1

void *Disp(void *) {

    while(true) {
      sem_wait(disp_sem);
      if (disp_data.type=TEMP){
        printf("Temperature: f", disp_data.value);
        if(disp_data.alarm_cond==ON)
          Start_T_Alarm();
        sem_post(temp_sem);
      }
      else {
        printf("Humidity: f", disp_data.value);
        if(disp_data.alarm_cond==ON)
          Start_H_Alarm();
        sem_post(humid_sem);
      }
}
```

The main program will initialize data structures including the semaphores and activate these threads which we do not show.

11.5 Ada

The Ada language was developed as a request of the US Department of Defense to have a programming language for mission-critical systems. It is a widely used language with support for concurrency and real time. It was published as an ANSI/MIL standard in 1983 and as an ISO standard in 1987 [1]. The second major revision of Ada resulted in Ada 95 which we will briefly review. Ada is block-structured and consists of one or more of the following modules:

- **subprogram**: It is like the procedure or function in a general programming language.
- **package**: It is used for encapsulation and modular design.
- **task**: A task is the basic unit of concurrency representing the operating system task.
- **protected unit**: It is mainly used for shared data with added synchronization.

The structure of a block in Ada has the following structure [2].

```
declare
-- definitions of objects, subprograms, types, etc.
begin
-- sequence of statements
exception
-- exception handling
end;
```

Let us see how we can implement a function that finds the sum of the elements of a vector of integers in Ada. The *Summation* function returns the *Sum* variable; note that *V'Range* is all we need to do to have the loop index *I* run through all values of the vector *V*.

```
function Summation(V: Vector) return Integer is
   Sum : Integer := 0;
begin
  for I in V'Range loop
      Sum := Sum + V(I);
  end loop;
  return Sum;
end Summation;
```

Ada has the basic characteristics of an object-oriented language, and a module can declare some of its variables and functions as *private* so as not to be accessible externally. Also, object-oriented programming facilities such as inheritance, constructors, destructors, and polymorphism are available in Ada 95.

11.5.1 Concurrency

The basic unit of concurrency in Ada is called a *task* which is declared explicitly using the keyword *task*. Tasks synchronize and communicate as the operating system tasks using shared variables, protected units, or a method called *rendezvous*. A task has a specification that contains its name, its possible input parameters at creation, a visible part, and a private part. The body of a task contains its executable code as shown in the example below.

```
procedure Example1 is
task type A_Type;
task B;
A,C : A_Type;
task body A_Type is
--local declarations for task A and C
begin
--sequence of statements for task A and C
end A_Type;
task body B is
--local declarations for task B
begin
--sequence of statements for task B
end B;
begin
--task A,C and B start their executions before the first
-- statement of this procedure.
end
```

In this program segment, we define a task type *A_type* which can be used to classify tasks. The task body declaration contains local declarations and the actual code for the task. When the procedure begins, all tasks start execution concurrently.

11.5.1.1 Time Management

A data type called *Time* represents the real time with a resolution of 1ms in Ada 95. The value of current real time can be read by the function *Clock*. The following example demonstrates how to measure the time that the execution of a loop takes.

```
task body Example2 is
  First, Second, Interval : Time;
  begin
    First := Clock;     -- record the time before loop starts
    loop
    ...
    end loop;
    Second := Clock;    -- record the time when loop ends
    Interval := Second - First;
  end Example2;
```

Not that we could have used "Interval := Clock − First," statement instead of the two statements after the loop ending. Alternatively, Ada provides the use of the package *Calendar* to manage time effectively.

11.5.1.2 Periodic Tasks

A periodic task in Ada can be specified using the *delay until* structure as shown in the example below. The size of the interval may be set to reflect the period. Note that *Interval* should be set to $T - C$ for the task, for example, if task executes 20 ms and should be activated every 100 ms, we need to set the *Interval* value to 80 ms.

```
task  T1 is
Interval : constant Duration := 0.08;
Next_Time : Time;
begin
  Next_Time := Clock + Interval;
  loop
     Action;    -- procedure doing useful work
     delay until Next_Time;
     Next_Time := Next_Time + Interval;
  end loop;
end T;
```

11.5.1.3 Task Priorities

Support for static and dynamic task priorities is provided in Ada. The static priority of a task is specified using the pragma *Priority* which is contained in the specification of the task as shown below.

```
task  T1 is
  #pragma  Priority(12)
end T;
```

The real-time systems addition to Ada 95 supports dynamic priorities.

```
package Ada.Dynamic_Priorities is
  procedure Set_Priority(...);
  function Get_Priority(...) return Any_Priority;
end Ada.Dynamic_Priorities;
```

Using this package, the priority of a task can be read and changed during its execution.

11.5.1.4 Task Synchronization and Communication

The main synchronization and communication method in Ada is the use of the pair
entry/accept calls. Two tasks, a *caller* and a *server*, synchronize and transfer data by
these calls. The caller calls an *entry* in the server, and the server executes the *accept*
statement to receive the call. If execution of *accept* is made prior to the *entry* call
by the caller, the server is blocked at accept point. Otherwise, if the entry execution
is made prior to the server execution of *accept*, the caller is blocked at its entry
execution point. The caller remains blocked during the processing of the *accept* by
the server. This way, a full and reliable synchronous communication is achieved.
The caller needs to know the server and its entries but the server does not need to
know the callers. Only one caller may rendezvous with a server, and any other callers
attempting to rendezvous are blocked. The following example illustrates a server that
accepts an integer from a client, squares it, and prints the output. For example, the
client needs to make the call *Square.Calculate(3)* to find the square of 3. The output
is achieved by the *Put* call.

```ada
task Server is
    entry Square(x in Integer);
end Server;

task body Server is
    a : Integer;
begin
    accept Square(x : in Integer, a : out integer) do
        a := x * x;
        Put(a);
    end Square;
end Server;
```

When a server wants to wait on two or more different types of *accept* calls, the
select statement may be used. The following example illustrates the use of select
where the server task may accept *Add* or *Subtract* entries in any order and blocks
until an accepted call is finished.

```ada
task Calculate is
    entry Add(x,y in Integer, z: out Integer);
    entry Subtract(x,y in Integer, z: out Integer););
end Calculate;

task body Calculate is
    a : Integer;
begin
```

```
      loop
         select
            accept Add(x, y: in Integer, a: out Integer) do
               a:= x + y;
               Put(a);
            end Add;
         or
            accept Subtract(x, y:in Integer, a:out Integer)do
               a:= x - y;
               Put(a);
            end Subtract;
      end Calculate;
```

11.5.2 Exception Handling

Exceptions in Ada language are the error conditions that arise during program execution. *Exception handling* is the process of catching these errors at runtime and performing some remedy action or displaying error status. A block statement has a declarative part, a sequence of statements, and an exception handling section in Ada. The predefined exception types in Ada are defined in package *Standard* as follows.

- *Constraint_Error*: Raised when a range constraint is violated.
- *Numeric_Error*: Raised when a numeric operation cannot be performed due to conditions such as overflow or divide by zero.
- *Program_Error*: Raised when the end of a function is reached without encountering a *return* statement.
- *Storage_Error*: Raised when memory space is exhausted due to a dynamic creation of an object or due to exhaustion of stack space.
- *Tasking_Error*: Raised during task communications, for example, a task trying to synchronize by a dormant task using rendezvous.

Exceptions can be handled by the *exception* part at the end of a block using the *when* keyword to specify the type of exception. Let us illustrate these concepts by a recursive function that computes factorial of an input integer. If the input value is too large that can be computed, an exception is raised.

```
function Factorial(n:integer) return integer is
begin
  if n=1 then return 1;
  else return n * Factorial(n-1);
  end if;
```

```
exception
 when NUMERIC_ERROR
     Put_Line("Input too large");
end
```

We can have more than one exception handling in a block, for example that is used to open and perform some operation on a file, as shown below. Exception handling is performed similar to the *case* statement with each case containing one exception handler.

```
begin
 -- operations on the file
exception
 when File_Not_Found
     Put_Line("File does not exist");
   when End_Of_File
     Close(file);
   when others
     Put_Line("Error");
end
```

An exception can be activated by the user using the call **raise** and then specifying the exception type anywhere in the program as in the example below.

```
        raise CONSTRAINT_ERROR
```

Users can define their own exceptions by inserting them at the declaration part as follows.

```
        my_exception: exception;
```

Exceptions can be suppressed by the use of pragma *suppress*. For example, the syntax for suppressing STORAGE_ERROR is as follows.

```
        pragma suppress (storage_check);
```

11.5.3   Ada Process Control Implementation

The implementation of real-time software for the process control system of Fig. 11.1 involves three Ada tasks as in the C/POSIX implementation. Mutual exclusion between tasks *Temp* and *Humid* can be provided by the *protected* object *Display*. Any procedure in this object will be executed by one task only, and a calling task will be blocked if any procedure is executed by another task.

```
protected type Display is
  procedure Temp_Disp(data: in Temp_Data);
  procedure Humid_Disp(data: in Humid_Data);
end Display;
protected body Display is
begin
  procedure Temp_Disp(data : in Temp_Data) is
    begin
        Printline("Temperature: ", data);
    end Temp_Disp;
  procedure Humid_Disp(data : in Humid_Data) is
    begin
        Printline("Humidity: ", data);
    end Humid_Disp;
end Display;
```

The two tasks can now be coded as follows: each task, after reading the value from its sensor, checks whether this value is below or higher than threshold values and calls its procedure in the object *Display*.

```
procedure Process_Control is
    task Temp;
    task Humid;
    task body Temp is
        temp_Uthreshold: constant := (some_value);
        temp_Lthreshold: constant := (some_value);
    begin
      loop
        Read_Temp(temp_value);
        if temp_value > temp_Uthreshold or
         temp_value < Ltemp_threshold  then
            Set_Temp(Temp_Switch, temp_value);
        end if
        Display.Temp_Disp(temp_value);
      end loop
    end Temp;
    task body Humid is
    begin
        Humid_Hthreshold: constant := (some_value);
        Humid_Lthreshold: constant := (some_value);
        loop
            Read_Humid(humid_value);
            if humid_value > Humid_Uthreshold or
                temp_value < Humid_Lthreshold  then
                Set_Humid(Humid_Switch, humid_value);
```

```
        end if
        Display.Humid_Disp(humid_value);
      end loop
    end Humid;

    begin
      null; -- Temp and Humid tasks run concurrently
    end Process_Control;
```

11.6 Java

Java is a platform-independent object-oriented language with a similar syntax to C/C++. One of the main goals of Java is to be portable and for this reason, Java source code is compiled to an intermediate code called *Java bytecode* which can run on any *Java virtual machine* (JVM). It exhibits all of the basic features of object-oriented programming such as classes, inheritance, and polymorphism. Let us implement the previous function of summing of elements of a vector in Java. We first define a class *Vectors* and then a method in this class called *Summation* that calculates the sum of the entries of an input vector.

```
class Vectors
{ public int  Summation(Vector V )
  { int sum = 0;
    int i;
    for (i=0; i<V.length; i++)
      sum = sum + V[i];
    return sum;
  }
}
```

A class in Java contains data and methods and a class may belong to a *package*. Methods and instance variables in Java can be *public* which are visible outside the class; *protected* that are visible only within package or in a class, or *private* that may not be accessed outside the class.

11.6.1 Java Threads

A thread in Java is the basic unit of execution as a POSIX thread or an Ada task. The JVM main thread starts by calling the method *main*. The class *java.lang.Thread* can be extended to create threads. There is also the standard interface which can be used for concurrent execution as below.

```
public interface Runnable
    public abstract void run();
```

As the first thread creation example, we use the *Thread* class in the following example where this class is extended to *thr* class and the instance *T1* of this class is created by the system call *new* in the main thread. The created thread simply writes "Hello" on the screen. The code specific to the thread can be placed inside the *run* method in this implementation.

```
class thr extends Thread {
    public void run() {
        System.out.println("Hello");
    }
    public static void main(String args[]) {
        thr T1= new thr();
        T1.start;
    }
```

The second type of implementation involves using the *Runnable* interface and starts with the "**class** thr **implements** Runnable" heading to specify the *run* method. Dynamic creation of threads is possible and the main program terminates when all of the user threads have terminated. Java threads can either be *user threads* or server threads called the *daemon threads*. Commonly used methods with Java threads are as follows.

- *void run()*: Entry point for the thread.
- *void start()*: Start a thread by activating its run method.
- *boolean isAlive()*: Determines if a thread is still running.
- *sleep(long ms)*: Suspend the thread for the specified period of time.
- *void join()*: Wait for a thread to finish.

11.6.2 Thread Synchronization

A critical section in Java is protected by the *lock* objects. Two methods defined on a lock object are *lock* and *unlock* as shown in the below code segment.

```
import java.util.concurrent.locks.*;
// non-critical section
Lock my_lock;
while (true) {
  my_lock.lock(); // enter critical section
  try {
  // critical section
}
finally { // this is needed to ensure releasing
// of the lock even when an exception is raised
my_lock.unlock(); // return to normal mode
  }
}
```

The public *semaphore* class in Java has the following methods that can be implemented.

- *count( )*: returns the current value of the semaphore.
- *up( )*: increments the semaphore count. This call is analogous to *signal_sema* call in DRTK.
- *down( )*: waits until count is positive and then decrements the count. This call is analogous to *wait_sema* call in DRTK.

Semaphores can be used for inter-thread synchronization and also for mutual exclusion as the locks. Assuming a semaphore object *sem* is created, a critical section can then be protected by the *sem.down* and *sem.up* calls upon entrance and exit, respectively.

11.6.2.1   Management of Time and Scheduling

Java provides the public class *Clock* and *getTime* and other methods on this class for time management. A thread in Java can be interrupted by another thread using the function *interrupt( )* which raises an *InterruptedException*. A sleeping or a waiting thread can catch this interrupt; however, a running thread will not catch the interrupt. The function *interrupted( )* which returns boolean true when an interrupt occurs may be checked by the thread to catch the interrupt as in the following code segment. The periodic thread sleeps for 200 milliseconds before the next activation. However, it may receive interrupts and serve them.

```
class thr extends Thread {
   public void run() {
       try {
           for(;;) {
               while(!interrupted) {
                   // do some work
                   Thread.sleep(200);
               }
           }
       } catch(InterruptedException e) {
           // do interrupt serving
       }
   }
}
```

11.6.3 Exception Handling

Exception handling in Java is achieved by the *try-block* which provides guarding of a block.

```
try
// code that may cause exception
// ...

catch  (exception type x)
   // exception handler for x

finally  (exception type y)
   // code executed in all cases
```

The following types of exceptions in Java are possible.

- The *NullPointerException* is thrown when a null pointer is passed to the stop method.
- The *IllegalThreadStateException* is thrown when the start method is called and the thread has already been started.
- The *SecurityException* is thrown when a stop or destroy method has been called on a thread for which the caller does not have the permissions for the operation requested.

- The *InterruptedException* is thrown when a thread is waiting or sleeping, and it is interrupted either before or during this activity.

11.7 Review Questions

1. What are the main requirements from any real-time programming language?
2. How does C programming language provide concurrency?
3. What is the main function in UNIX to delay a task? What is the function to delay a task in POSIX interface?
4. What is the main system call in Ada to make a periodic task?
5. What is the main method of task synchronization in Ada?
6. What are the main exception types in Ada?
7. What is the main characteristic of Java that makes it different from many other programming languages?
8. What are the two main methods of thread creation in Java?
9. How is mutual exclusion when accessing a shared variable is achieved in Java?
10. What are the main methods of the semaphore class in Java?
11. How is an exception caught in Java?
12. What are the main exception types in Java?

11.8 Chapter Notes

The main requirements from a real-time programming language are efficient time management, support for concurrency, facilities for exception, and interrupt handling and real-time scheduling support. We reviewed three main real-time programming languages with these issues in mind: C/POSIX, Ada, and Java. The C language is still widely used to implement real-time software but itself is not real time; however, real-time POSIX extension provides convenient management of time, multitasking using threads, and timed task synchronization and communication primitives. Ada is designed for mission-critical systems and has all of the facilities stated above for real-time programming. A task can be defined as an individual executable unit and various methods for inter-task synchronization and communication are provided. Ada has facilities for exception handling and scheduling. The Java programming language can be used for software implementation in real-time systems as it provides concurrency by threads and has facilities for interrupt and exception handling.

The fundamental decision is whether to use primarily a real-time operating system or a real-time programming language to implement real-time software. The former means we can use a general programming language with real-time operating system interface, for example, C/C++, with ease in portability and possibly with a richer library of system calls than a real-time programming language. On the other hand, using a real-time programming language may be complicated if the models used by the language and the operating system are radically different. We have indicated a very brief introduction of the real-time features of the three languages mentioned: real-time C/POSIX is described in [2], a thorough study of Ada can be found in [3], and use of Java for real-time applications is in [4].

11.9 Programming Exercises

1. Write a program in C/POSIX with three threads $T1$, $T2$, and $T3$, where $T1$ inputs an integer from the user and writes it to a shared memory place. The $T2$ thread reads this integer, checks it against a lower and a higher limit, multiplies it with a constant, and places it in another shared memory place so that $T3$ inputs the value and displays it. All critical sections should be protected.
2. Write a program in Ada with a server task that acts as a calculator to perform the four basic arithmetic operations and a caller task that receives input from the user as two numbers and the needed operation. The caller calls entry of the needed accepting part of the server.
3. Write a program in Ada that declares an array X of 10 integers and initializes each entry so that $X(I) = 2 * I$. The program in execution inputs an index from the user and displays the entry of X in that index. Provide the exception handling part of the program to have the CONSTRAINT_ERROR raised and an error message displayed when the user enters an invalid index.
4. A Java module provides periodic real-time tasks $T1$, $T2$, and $T3$. Task $T1$ reads data from a file and then sends an interrupt to task $T2$ which writes this data to another file and performs an up on a semaphore which task $T3$ is waiting. Task $T3$ reads the contents of file and displays it. Write this program in Java with short comments.

References

1. Ben-Ari M (2005) Ada for software engineers. Weizmann Institute of Science
2. Burns A, Wellings A (2001) Real time systems and programming languages: Ada 95, real-time Java and real-time C/POSIX, 3rd edn. Addison-Wesley
3. Burns A, Wellings A (2007) Concurrent and real-time programming in Ada, 3rd edn. Cambridge University Press
4. Real-time specification for Java 2.0 (RTSJ 2.0). https://www.aicas.com/cms/en/rtsj

Fault Tolerance

12

12.1 Introduction

A fault in a computer system is manifested by the malfunctioning of a component related to the hardware or the software of the system. Component defects may arise because of manufacturing faults or due to aging, that is, wear and tear effect due to long-time use or some environmental effects such as using the component out of range of its reported temperature or vibration range, etc. A software fault can arise due to various causes such as software bugs, erroneous design, etc. A fault causes an error which may result in the failure of a part or the whole of the system. Tragic disasters in the past such as airplane crashes, rocket falls, etc. follow this pattern commonly. For example, a fault in the circuit component of the altitude display of an airplane will cause error in the reading of the display which may direct the pilot wrongly to cause an accident.

A fault needs to be detected, and recovery action from the fault is then to be invoked to prevent a failure if this is possible. Fault tolerance is the ability of the system to continue functioning correctly in the presence of faults. Fault tolerance is imperative in a real-time system since a resulting failure may cause loss of lives and property.

We start this chapter with the basic concepts and terminology related to fault tolerance in general. We then look at basic hardware and software fault examples, review methods for fault recovery, and describe fault-tolerant scheduling in real-time systems. Employing task groups is a fundamental method for fault tolerance in distributed (real-time) systems and we discuss methods for orderly and reliable delivery of messages to a task group for this purpose.

© Springer Nature Switzerland AG 2019
K. Erciyes, *Distributed Real-Time Systems*, Computer Communications
and Networks, https://doi.org/10.1007/978-3-030-22570-4_12

12.2 Concepts and Terminology

Faults in a computer system arise due to many reasons with the fundamental ones being inadequate specification, software design errors, hardware component failures, and networks errors mainly due to interference. An active fault is the cause of an error which is a partial system state that may result in system failure. An error may propagate causing a series of errors. We need to define several terms for fault tolerance as below [8,10].

- *Dependability*: This is the ability of a system to deliver its intended level of service to its users [10]. The main characteristics of dependability are reliability, availability, and safety.
- *Reliability, Unreliability*: Reliability $R(t)$ is the probability that the system functions according to its specifications in the interval $[0, t]$, given it was functioning correctly at time 0. Unreliability $F(t)$ is the probability that the system fails at any time in the interval $[0, t]$. It follows that $R(t) = 1 - F(t)$.
- *Availability*: Availability $A(t)$ is the probability that the system is running correctly with respect to its specifications at time t. Note that availability is evaluated for a particular instant t, whereas reliability is evaluated for the interval $[0, t]$.
- *Safety*: Safety $S(t)$ is the probability that the system does not fail in the interval $[0, t]$ and will function correctly so as not to cause damage to people, property, or environment.
- *Maintainability* $M(\Delta t)$: This is the probability that a system with a failure will be restored within a specified interval of time Δt.
- *A Safety-Critical System*: This is a system, failure of which may result in damage to people, property, and/or environment. Distributed real-time and some embedded systems generally fall in this category.
- *Security*: This is the ability of the system to protect itself from a potential damage.
- *Failure, error, and fault*: A system designed has *specifications* stating its behavior. When the behavior of the system diverts from its specifications, the state of the system is a *failure* which is caused by an *error* resulting from a *fault*. Figure 12.1 depicts this relationship.
- *Error Latency*: This is the time between the occurrence of an error and the resulting failure.

Fault prevention is a set of methods to prevent occurrence of a fault. Two main approaches for handling faults are *fault masking* and *fault reconfiguration*. Fault masking is the process of preventing faults to occur in a system, whereas fault reconfiguration involves eliminating the faulty components from the system and restoring

Fig. 12.1 Relationship between fault, error, and failure

the system to an operational state which consists of the following consecutive steps in general.

1. *Fault Detection*: A fault needs to be detected before starting any recovery procedure.
2. *Fault Location*: This is finding the location of the fault.
3. *Fault Containment*: A fault needs to be isolated to prevent it from propagating in the system.
4. *Fault Recovery*: This is the process of restoring the system to an operational state after the occurrence of a fault.

12.3 Fault Classification

Faults may arise due to various reasons, for example, a specification error or an error due to external disturbances such as noise or radiation. Based on the occurrence and duration of a fault, it can be classified as below.

- *Permanent Faults*: A permanent fault typically is caused by a failing component, and recovery from such a fault in general is only possible by replacing or repairing that component.
- *Transient Faults*: A transient fault occurs for a specified time only, and the system generally continues functioning correctly afterward. These faults disappear after a short duration. Many communication network faults are of this type.
- *Intermittent Faults*: An intermittent fault is a reoccurring fault that is difficult to detect. The fault occurs and then the system operates correctly, and this situation occurs frequently. A loose connection in an electronic circuit may cause such a fault.

Viewed from another angle, faults can be classified with respect to the failures they produce as follows, considering the values produced and the times they are produced [1,6].

- *Fail-safe*: The system continues functioning correctly in the presence of temporary faults.
- *Fail-late*: Correct values are produced by the system but there are delays in the time these are produced.
- *Graceful degradation* or *Fail-soft*: The system continuous with its main functions while degrading some of the services when a fault occurs.
- *Fail-silent*: A system that works correctly in time and value domains until an omission (process or communication link) failure occurs after which omission failures in all services occur.

12.4 Redundancy

Redundancy is the duplication of critical hardware and/or software components of a system to increase reliability by incorporating the use of a duplicate when the active component fails. The main types of redundancy are the hardware redundancy, information redundancy, time redundancy, and the software redundancy.

12.4.1 Hardware Redundancy

Hardware redundancy in a computer system can be implemented as *passive redundancy*, *active redundancy*, or *hybrid redundancy* [5]. In passive hardware redundancy, fault masking is employed. Commonly, an *n* independent identical hardware modules perform the same function outputs of which are voted to decide on the correct output. An *M*-of-*N* system consists of *N* components, and the correct operation of this system requires at least *M* components working correctly. The triple modular redundancy (TMR) system is a 2-of-3 system with $M = 2$ and $N = 3$, needing two correctly operating components for the correct operation of the system. TMR is realized by three components performing the same action, and the result is voted. The voting circuit in a TMR implementation may be realized by AND–OR gates called *1-bit majority voter* as shown in Fig. 12.2. There are three modules *A*, *B*, and *C* which produce three 1-bit outputs *a*, *b*, and *c*, respectively. Let us assume faulty state produces a *false* output and a correct state produces a *true* output. Note that the output of the OR gate is *true* even if one component fails in this case. An *n*-bit majority voter is *n* of these 1-bit voters running in parallel. Total delay in such a voter is 2-gate delay by the AND gates and the OR gate.

TMR voting can be performed in software by a simple program that compares the outputs from the modules using three comparisons and outputting the one that is the same with at least the outputs of two modules. Hardware voter is more expensive

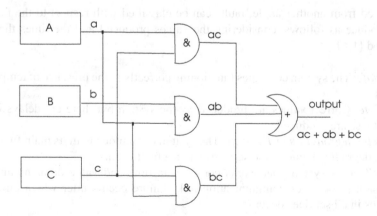

Fig. 12.2 A TMR realization with logical outputs

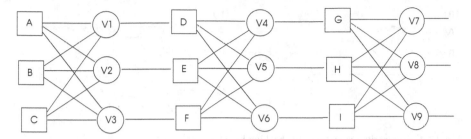

Fig. 12.3 A general TMR realization. Adapted from [13]

than the software voter because of the extra circuit needed but is faster. In a more general sense, the TMR with three stages is realized as shown in Fig. 12.3 without the need for the logical variables. Also, it is considered that voters can be faulty and hence three stages.

N-modular redundancy is the generalization of TMR with N modules. Faults up to k times can be detected when $N > 2k$. A *watchdog timer* is reset periodically by the system/application tasks and failure of doing so indicates a fault. Watchdog timers are commonly used in real-time systems. Active hardware redundancy is achieved by detecting the fault after which fault location, containment, and recovery procedures are performed. Hybrid hardware redundancy combines the passive and active redundancy approaches where fault prevention and steps of active redundancy are employed [5].

12.4.2 Information Redundancy

Data transferred from one memory location to another, or more commonly, from one node in a network to another may contain errors. *Information redundancy* is adding more information to data to detect errors.

12.4.2.1 Coding

This operation at data word level is known as *coding* in which check bits are added to data. A d-bit word is encoded to a c-bit *codeword*, and the receiver/user of the data *decodes* the codeword to retrieve the original data [8]. A coding scheme can be designed so that errors introduced in a codeword cause out-of-range codewords and hence, errors can be detected. Coding can also be designed to enable correction of the codeword from the erroneous one in *error correction*.

Hamming distance of two binary words x and y is the number of bits they differ. For example, given two binary words $x = 0100\ 1000$ and $y = 0101\ 1010$, Hamming distance of x and y, $H_d(x, y)$, is 2.

(a)

(b)

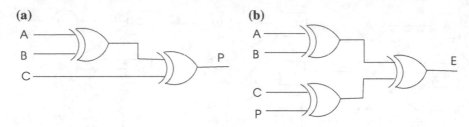

Fig. 12.4 Even parity realization and checking by XOR gates

Table 12.1 An example codeword matrix of horizontal sand vertical even parity

c_4	c_3	c_2	c_1	c_0
1	0	1	0	**0**
0	1	1	1	**1**
1	0	0	1	**0**
1	1	1	1	**0**
1	**0**	**1**	**1**	**1**

Parity Codes

An extra bit can be added to a binary word so that the resulting word has even (*even parity*) or odd (*odd parity*) number of 1s. For example, given the binary 7-bit word 0110 101, we need to add a 0 for even parity or a 1 for odd parity to it as the last bit. The receiver of the codeword can then test the parity and detect a 1-bit error. Simplicity of this method allows its use in protection of data in memory. Parity generation and detection is commonly performed in hardware using XOR gates as shown in Fig. 12.4 where a parity bit for a 3-bit data using even parity is generated in (a) and a 4-bit codeword is tested for even parity to output an error bit (E) in (b).

In order to detect more than 1-bit errors, horizontal and vertical parity generation and testing can be employed. This method adds horizontal parity bits as before and additionally adds vertical parity bits to form a new codeword at the bottom. This way, a single bit error in a block of data can be corrected. Such an erroneous bit in the codeword would have a parity error on both the row and the column it belongs. An example of horizontal and vertical parity is shown in Table 12.1 where four 4-bit data words (c_4 to c_1) with an added even parity bit (c_0) are vertically added to form the fifth word with all added parity bits shown in bold.

Cyclic Redundancy Check

Cyclic Redundancy Check (CRC) is an efficient method for error detection used commonly in the data link layer of OSI seven-layer model. A CRC node calculates a short binary code called *check value* and appends it to the data to be sent/written to form a codeword. This is done by the following steps applied to n bits of binary data $D(n) = \{d_{n-1}, \ldots, d_0\}$.

Fig. 12.5 A CRC example

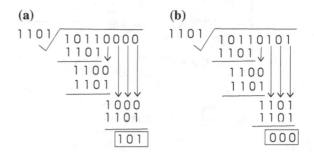

1. Select a polynomial word $P(k)$ of size k.
2. Append $k - 1$ 0s to the end of $D(n)$ to get $D(n + k - 1)$.
3. Divide $D(n + k - 1)$ by $P(k)$ using modulo-2 arithmetic. This is like binary addition except carry bits are discarded.
4. Append the remainder $R(k - 1)$ obtained to the end of $D(n)$ to get $D^*(n + k * 1)$ and send/write it.

The receiver/reader does the following steps.

1. Divide the received/read $D^*(n + k - 1)$ by $P(k)$ using modulo-2 arithmetic.
2. If the remainder is zero, the received message does not contain errors, submit $D^*(n)$ to the application.
3. Else notify error.

An example of the sending side and receiving side is shown in Fig. 12.5 where $D(5) = 10110$ and $P(4) = 1101$. The remainder obtained from mod-2 division in (a) is 101 which is appended to $D(5)$ to get 10110101 which is transmitted/written. The receiver/reader then divides this binary data to the same polynomial and gets zero in (b) which means no errors have occurred in this case.

It can be shown that the remainder at the receiver will always be zero if there are no errors using this method. Choice of $P(k)$ is important as it should maximize error-detecting capabilities and minimize collision probabilities. Most commonly used CRC polynomial lengths are 9, 17, 33, and 65 bits.

12.4.3 Time Redundancy

Time redundancy is achieved by allocating extra time to check faults, for example, repetitive execution of a software module and comparing the results will provide detection of transient faults as these faults typically occur only few times throughout system lifetime. In Fig. 12.6, the logical diagram of time redundancy is depicted where the same computation is repeated with the same input x at three time points; t_1, t_2, and t_3 and the outputs a, b, and c are compared. If one of the first two outputs are different than c, a transient fault has occurred. Note that if c is different than both

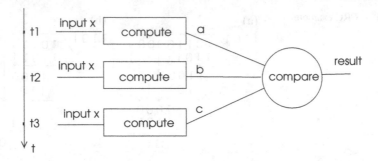

Fig. 12.6 Time redundancy

a and *b*, we need to perform another test at another time point to decide whether fault at time t_3 is permanent or transient. This type of redundancy does not need any extra hardware but may result in high computation overhead.

12.4.4   Software Redundancy

A software fault is commonly the result of a faulty design which occurs due to a specific input sequence or condition. Software redundancy methods are based on using extra software storage and running to prevent faults occurring. These fault-tolerant techniques may be broadly classified as *single-version* and *multi-version* methods.

12.4.4.1   Single-Version Methods

These methods employ additional software modules to the single version of the software module to achieve fault recovery. Two techniques for fault recovery are *exception handling* and *checkpointing*.

Exception Handling

Exceptions are software interrupts due to some unexpected condition such as divide by zero, memory corruption, and also an erroneous design effect. Proper handling of exceptions is needed as they may result in system failures. Exception handling methods can be broadly classified as *programmed exception handling* and *default exception handling*. The type of exceptions that can occur is known in the former enabling the action to be taken by the programmer. An exception handler will attempt to mask the fault and if this is not possible, the system may be brought to a previous safe state by checkpointing recovery. The default exception handlers should deal with all unpredicted exceptions by the use of programming language constructs. Exception handling is specified in various programming languages such as ADA and Java as we reviewed in Chap. 11. Handling of exceptions by default is commonly performed by the programmer using an existing programming language construct.

Exception handling is important in real-time systems as failure in these systems may have catastrophic results. An exception in such a system will cause computational overhead which may result in missed deadlines of hard real-time tasks. Hence, exception handling code in a real-time system should be very effective.

Checkpointing

Checkpointing method is based on restoring system state to a previous fault-free state when failure occurs. Current state of the system is saved on a nonvolatile storage periodically or before critical code executions. Only the modified system state is saved in *incremental checkpointing* saving time and space during storing the system state. Detecting the failure is part of the checkpointing software which is commonly initiated by a condition such as detecting a terminating task which should work continuously. Restoration of the system is performed by loading from the last recorded state and then continuing with the normal operation. In the case of incremental checkpointing, the last full checkpointing state is restored and then the incremental changes are implemented.

This method will be able to detect transient faults but the permanent failures will continue resulting in more restoring processes. An error that did not show up may be stored with the system state, and restoring system state will not eliminate the error. In such cases, system reboot or reporting the failure can be performed. Graceful degradation using a different and possibly less efficient software module can also be performed in such a case.

12.4.4.2 Multi-version Redundancy

Multi-version methods employ multiple versions of the software component for redundancy to achieve fault recovery. Two main approaches in this technique used are the *recovery blocks* and *N-version programming*.

Recovery Blocks

The software is divided into blocks, and each block has a number of functionally equivalent blocks running in parallel with the primary block in this method. The entry point to a block is the *automatic recovery point*, and at the output an *acceptance test point* is formed. After executing the primary block, acceptance test is executed to test whether the system is in acceptable state. If this test fails, execution is diverted to the automatic recovery point at the beginning of the blocks and another block is selected for execution. Failing again results in execution of another parallel block and if all blocks fail, failure is reported and recovery process should be handled at a higher level of abstraction.

***N*-version Programming**

Software redundancy can be accomplished by *N*-version programming. The main idea of this technique is to test different versions of a program and test whether the outputs are the same to find the faulty component [3]. This method is analogous to NMR in hardware. The functional equivalent of a program is called *versions*.

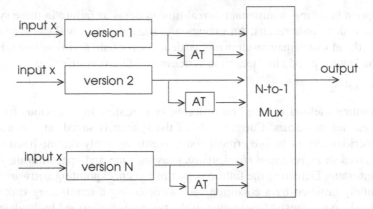

Fig. 12.7 N-self-checking programming

The N different versions of the program are executed in parallel, and the outputs are compared by a *driver* process in this method. When the initial conditions are the same and the inputs are the same for all versions and the decision algorithm is reliable, all outputs from the versions should be the same. The versions are commonly developed by different teams that do not interact during development. Different algorithms, programming languages, and development environments are encouraged to result in substantially different designs [12]. Sequential executions of the versions are also possible but not suitable for real-time applications.

N-Self-Checking Programming

The N-self-checking programming method uses both N-version programming and recovery blocks. There are N versions of the software module, and there exists an acceptance test (AT) at the output of each module. The highest ranking version that passes the acceptance test is taken as the output from the N-to-1 multiplexer as shown in Fig. 12.7 [5]. An N-self-checking programming using comparison compares outputs from each pair of versions and selects the highest ranking stable input to the multiplexer.

12.5 Fault-Tolerant Real-Time Systems

Fault tolerance is imperative in a real-time system since a failure may result in catastrophic events. All of the redundancy techniques described which are hardware, information, software, and time redundancies described are applicable to such systems. In reality, hardware redundancy is frequently employed in safety-critical systems such as an airplane. A fundamental requirement from a real-time system is to have task meet their deadlines. Hence, providing fault-tolerant scheduling techniques in the presence of faults is needed for this purpose, in addition to the general fault-tolerant methods. We have reviewed in Chap. 7 static scheduling algorithms which work offline and dynamic scheduling algorithms that decide the next task to schedule at runtime. Analysis of fault-tolerant real-time scheduling is described next.

12.5.1 Static Scheduling

Fault-tolerant static scheduling methods commonly reserve slack time in the schedule computed offline so that this available time can be used online when there is a fault in the system. Three main approaches used in fault-tolerant scheduling are primary/contingency scheduling, masking, and online rescheduling as described in [7].

In primary/contingency scheduling approach, there exists a primary schedule and precomputed contingency schedules. A contingency schedule is activated when a failure occurs to have the tasks meet their deadlines. Masking method employs n versions of a task always running on a set of computation elements which is basically a software redundancy method. Online rescheduling is achieved in two steps; a static schedule is first computed offline. A failure of a task generates an aperiodic request upon which some of the periodic tasks are reordered without violating their deadlines and a corresponding backup for the failure is then activated in the slack time.

12.5.2 Dynamic Scheduling

We will review extensions to rate monotonic (RM) algorithm to make it fault-tolerant. The RM algorithm assumed independent periodic tasks and assigned priorities to them based on the length of their periods. If the utilization of the task set $U = \sum_{i=1}^{n} C_i / T_i \leq n(2^n - 1)$, where C_i is the computation time and T_i is the period of task τ_i, then RM algorithm provides a feasible schedule.

RM algorithm is extended to consider fault tolerance in [7,11] where a task has multiple versions, each of which is assigned to a different processor. The bin-packing algorithm with first-fit heuristic is used to allocate task versions as shown in Algorithm 12.1. We have a task set $T = \{\tau_1, \ldots, \tau_n\}$ and each task τ_j has r versions; $\tau_j^1, \tau_j^2, \ldots, \tau_j^r$. At each iteration of the algorithm, a test is done to see whether the version τ_j^i under consideration can be RM scheduled with the already assigned tasks at processor P_k. If this test fails, the number of processors is incremented (line 7). If this number is more than available processors, then it is restored to the maximum possible number. For each new task to consider, the processor number is set to 1 (line 13).

Algorithm 12.1 *FT_RMSched*

1: **Input**: set $\mathcal{T} = \{\tau_1, ..., \tau_n\}$ of n periodic tasks
2: **Output**: Schedule of $\mathcal{T}$
3: **for all** $\tau_i \in \mathcal{T}$ **do**
4:     **while** $\exists$ unassigned τ_j^i **do**
5:         **if** $\exists$ an RM schedule for $\tau_j^i \cup$ {tasks already assigned to P_k} and $\nexists \tau_j^w$ for any w assigned to P_k **then**
6:            **assign** τ_j^i to processor P_k
7:         **else** $k \leftarrow k + 1$
8:         **end if**
9:         **if** $k > m$ **then**
10:           $m \leftarrow k$
11:         **end if**
12:     **end while**
13:     $k \leftarrow 1$
14: **end for**

12.6   Fault Tolerance in Distributed Real-Time Systems

The hardware redundancy method of duplicating components and voting is frequently employed in distributed real-time and embedded systems such as an airplane. Software redundancy in terms of task groups is also commonly exercised in these systems. We first define failure modes in DRTSs and describe how task groups can provide fault tolerance.

12.6.1   Failure Classification

Failures are the results of errors as noted. Task failures in a distributed system may be classified as follows [4,13].

- *Crash Failure*: A task that crashes stops functioning forever.
- *Omission Failure*: A task failing by omission fails to reply sending and receiving messages.
- *Timing Failure*: A server does not respond in the required time interval.
- *Response Failure*: Response from the server is not correct. In value failure, the value of the response is not correct and in state transition failure the server departs from the correct flow of control.
- *Byzantine Failure*: A Byzantine task can get involved in any arbitrary behavior, it can duplicate messages, send unreal messages, etc.

A *failure detector* is an oracle that provides information about the state of tasks in the system. A failure detector may consist of distributed modules each associated with a task in the system [2].

12.6.2   Task Groups Revisited

We have reviewed task groups as a common middleware module in both non-real-time and real-time systems in Chap. 6. This module included system calls to create a group, to join/exit a group, and sending and receiving multicast messages with a multicast message sent to all members of a group. A task group can be flat with all members equal or hierarchical with a leader (or coordinator). Request for some service from the group is transferred to the coordinator of the group which manages the required function in the group in the latter. A group is called *closed* if only group members can send messages to it. A task outside the group may send messages to it in the *open* group. An *overlapping group* allows its members to be members of other groups, whereas a *nonoverlapping group* has all members belonging to it only. Group membership can be managed by a *group server* which keeps track of the created groups and the members of the group.

12.6.2.1   Reliable Multicast

Multicast communication involves sending a message to *all* members of a group as noted. We will make a distinction between *receipt* and *delivery* of a message. Receiving a message implies that the message has arrived at a node but not processed yet, whereas its delivery means it has reached to the upper middleware or the application layer.

Multicast communication may be achieved using various approaches. In the simplest case, the multicast message can be broadcast to all nodes of the network and then filtering at each node provides delivery to local group members only. The Ethernet protocol supports the broadcast delivery of messages. Multicast communication can also be achieved by multiple unicasts when the sender is aware of all of the group members as shown in Fig. 12.8a, at the expense of a large number of acknowledgement messages from each receiver. This type of multicast is called the *basic multicast (B-Multicast)* which is based on a reliable one-to-one *send* operation.

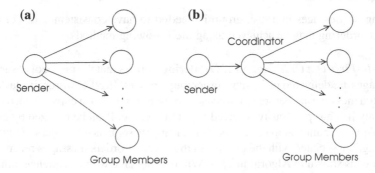

Fig. 12.8 Multicast modes

The hierarchical implementation is typically done by sending the message to the group coordinator (server) which distributes the message to all members of the group as in Fig. 12.8b. We have opted for this mode of operation in the DRTK implementation of task groups. A reliable multicast algorithm may be implemented using the B-Multicast algorithms as shown in Algorithm 12.2 where sending a message m to a group G and receiving a multicast message by a task τ are shown. The sender sends the message to itself and all members of the group using the basic multicast algorithm. Upon receiving the message, if the sender of the message is different from the identifier of the receiver, it is multicast to the group using the *B-Multicast* algorithm and then delivered the message to the middleware/application. A reliable multicast protocol does not perform ordering of messages, and an ordered delivery protocol may be implemented over the reliable delivery protocol.

Algorithm 12.2 *R-Multicast*

1: *received* ← ∅
2:
3: **procedure** SEND_R- MULTICAST(m, G)
4:     *B-Multicast(m,G)*
5: **end procedure**
6:
7: **procedure** RECEIVE_R- MULTICAST(m, G)
8:     **if** $m \neq received$ **then**
9:         *received* ← *received* ∪ {m}
10:         **if** $m.sender \neq my\_id$ **then**
11:             *B-Multicast(m,G)*
12:         **end if**
13:         *R-Deliver(m)*
14:     **end if**
15: **end procedure**

Message Ordering

Ordering of messages in a task group is needed to have consistent states of tasks. Message ordering may be achieved using the following methods.

- *First-In-First-Out Ordering*: A FIFO ordering implies that the protocol ensures the messages are delivered in the same order they are sent. In other words, if a task multicasts a message m_1 before it multicasts a message m_2, then a correct task receives m_2 only if it has previously received m_1. This protocol can be realized by assigning sequence numbers to messages. Sender inserts sequence numbers to multicast messages associated with the group and the receiver orders messages received out of order as shown in Algorithm 12.3. When a message with a sequence number k greater than the expected sequence number s arrives, it is placed in a delayed queue and delivered only when all the intermediate messages with sequence numbers in the interval $(k - s)$ are delivered.

Algorithm 12.3 *FIFO Ordering*

1: $seq\_no_G \leftarrow 0$
2:
3: **procedure** SEND_FIFO- MULTICAST(m, G)
4:     $m.seq\_no_G \leftarrow seq\_no_G$
5:     B-Multicast(m, G)
6:     $seq\_no_G \leftarrow seq\_no_G + 1$
7: **end procedure**
8:
9: **procedure** RECEIVE_FIFO- MULTICAST(m, G)
10:     **if** $m.seq\_no_G = seq\_no_G$ **then**
11:         FIFO-Deliver(m)
12:     **else if** $m.seq\_no_G > seq\_no_G$ **then**
13:         *insert* message into delayed message queue
14:         when all the messages in between are delivered, *FIFO-Deliver(m)*
15:     **end if**
16: **end procedure**

- *Casual Ordering*: *Causality* of tasks in a distributed system was specified by Lamport to mean the following [9]. Let precedence relation ($\rightarrow$) between events e_i and e_j mean e_i precedes e_j.

1. Events e_i and e_j are two events at the same node of the distributed system;
2. Event e_i is a sending of a message m and e_j is the reception of the message m;
3. There is an event e_k such that $e_i \rightarrow e_k$ and $e_k \rightarrow e_j$.

Casual-ordered multicast can be achieved using the vector clocks concept of time synchronization we have seen in Chap. 6 as shown in Algorithm 12.4 where p_i is the task identifier, the received message has p_j as the sender, and the vector clock of p_i is $V_i^G[N]$ for the group G. We need to ensure two conditions before delivering a message to its recipient; any message sent by p_j before the current message has been delivered and any message that was delivered by p_j has been delivered. These two checks at line 10 of the algorithm guarantee causal delivery, and the message is delivered only when these conditions hold.

- *Total Ordering*: This type of ordering guarantees all correct tasks that receive the multicast messages in the same order. A simple way to have total order multicast is to have a token that circulates the network in some predefined order. Any task that has the token may multicast using FIFO multicast.

12.7 DRTK Implementation

We will implement casual-ordered multicast using vector clocks with a group manager at each node used to send and receive multicast messages. The multicast message layout contains the vector clock values in the data field of the message as depicted

---

Algorithm 12.4 *Casual Ordering*

1: $V_i^G[N] \leftarrow 0$
2:
3: **procedure** SEND_CASUAL- MULTICAST(m, G)
4:     $m.V_i^G[j] \leftarrow V_i^G[j] + 1$
5:     *B-Multicast(m, G)*
6: **end procedure**
7:
8: **procedure** RECEIVE_CASUAL- MULTICAST(m, G)
9:     insert m in delayed message queue
10:     wait until $m.V_i^G[j] = V_i^G[j] + 1$ and $V_j^G[k] \leq V_i^G[k], k \neq j$
11:     *Casual-Deliver(m)*
12:     $V_i^G[j] = V_i^G[j] + 1$
13: **end procedure**

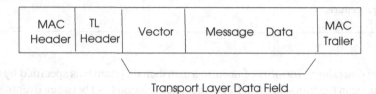

Transport Layer Data Field

Fig. 12.9 Multicast message format

in Fig. 12.9. We need to form a union structure in the message data field to be able to use this area for different middleware functions.

We will implement this protocol assuming multicast messages are deposited in the group mailbox rather than individually delivered to each group member. Note that the group manager mailbox and the group mailbox are two different places of storage, and all the group messages now have to go through group manager's mailbox. The data link layer input task (*DL_In*) described in Chap. 6 simply checks the type field in the message and puts the message in the mailbox of the group manager called causal order group manager (COGM) if it is a multicast message. This task continuously waits at its mailbox and when it receives a multicast message, it implements the vector clock rules to deliver it casually to the recipients via the group mailbox. Any message that is out of order should be delayed and queued. The group control block data structure of DRTK needs to be modified now; we have a vector data array and a data unit queue which are stored in the group control block of the group as shown below. Any delayed message after applying the causal order rule is placed in the data queue.

```
/************************************************************
                group data structure
*************************************************************/
/* group.h */
 #define ERR_GR_NONE    -2
 #define N_MEMBERS      30

typedef struct group  { ushort id;
                        ushort state;
                        ushort mailbox_id;
                        vector_t vector;
                        data_unit_queue data_que;
                        ushort n_members;
                        ushort local_members[N_MEMBERS];
                      }group_t;
typedef group_t* group_ptr_t;
```

We also need the *compare_vector* function to compare contents of two vectors using causal order rule as shown in the code below.

```
/************************************************************
            casual order data structures and functions
*************************************************************/
/* casual_order.c */

 #define WAIT_MSG   -1
 #define DELIVER     1

int compare_vec(ushort* local_pt, ushort* rem_pt){
   int i;
   for (i=0;i<N_GROUP_MEM;i++)
     if (*local_pt++>*rem_pt++)
       return(WAIT_MSG);
   return(DELIVER);
}
```

The COGM task can be implemented to consist of the following steps.

1. Receive a multicast message from the data link.
2. Check the contents of the vector clock in the incoming message m. Assume j is the task identifier of the sender and i is this node; if $m.V_i^G[j] = V_i^G[j] + 1$ and $V_j^G[k] \leq V_i^G[k], \forall k \neq j$, deliver the message. This is performed by storing the message in the mailbox of the group specified. Otherwise, the message is deposited in the vector table in the group control block of the specified group with its corresponding vector.
3. When a message is delivered, we need to check vector table entries in the group control block. Any message that obeys the casual order rule should also be delivered.

The above steps are applied in the COGM task shown below which queues the message in the vector table when the message needs to be delayed. It updates the vector in the group control block and then searches the data unit queue data vector fields to check if any backlogged message needs to be delivered now. If such a message is found, it is taken from the queue and delivered to the group mailbox. Details of *take_data_unit* function are not shown.

```
/*************************************************************
                 Casual Order Group Manager
*************************************************************/
TASK COGM() {

    data_unit_ptr_t recvd_pt, data_pt;
    group_ptr_t group_pt;
    ushort group_id, sender_id, mbox_id2;
    ushort mbox_id1=&(tcb_tab[current_pid])->mailbox_id;

    while(TRUE) {
    recvd_pt=recv_mailbox_wait(mbox_id1);
     group_id=(recvd_pt->TL_header).receiver_id;
     group_pt=&(group_tab[group_id]);
    mbox_id2=group_tab[group_id].mailbox_id;
    sender_id=data_pt->MAC_header.sender_id;
    if (recvd_pt->data.vector[sender_id]==
         group_pt->vector[sender_id]+1) &&
        (compare_vec(data_pt->data.vector,group_pt->vector)) {
      send_mbox_notwait(mbox_id2,data_pt);
      group_pt->vector[sender_id]++;
      data_pt=(group_pt->data_que).front;
      while (data_pt->next!=NULL)
       if (compare_vec(recvd_pt->data.vector,
              data_pt->data.vector)) {
        data_pt=take_data_unit(group_pt->data_que);
        group_pt->vector[data_pt->sender_id]++;
        send_mbox(mbox_id2,data_pt);
       }
     }
      else
        enqueue_data_unit(group_pt->data_que, recvd_pt);
    }
}
```

Sending a multicast message is similar to the *send_msg_notwait* procedure of DRTK with the exception that we need to increment the message count for the receiver and store it in the message before sending as follows, assuming *data_pt* is the address of the message. The rest of the code which is the same as *send_msg_notwait* is not shown.

```
vector_tab[current_tid].V[tid]=vector_tab[current_tid].V[tid]+1;
data_pt->data.V[tid]= vector_tab[current_tid].V[tid];
```

12.8 Review Questions

1. What is the relationship between a fault, an error, and a failure?
2. What is a dependable system?
3. What is the relationship between reliability, unreliability, and failure?
4. How are faults classified?
5. What are the main types of redundancy methods?
6. What are the main types of software redundancy?
7. What is checkpointing?
8. How does a recovery block method of software redundancy work?
9. What is the difference between N-version programming and N-self-checking programming?
10. What are the main methods used in fault-tolerant static real-time scheduling?
11. How is the RM scheduling modified to allow fault-free operation in a distributed real-time system?
12. What are the main types of failure in a distributed (real-time) system?
13. How is a task group used for fault tolerance?
14. What is reliable multicast?
15. What is the relation between the vector clocks and casual ordering?

12.9 Chapter Notes

We reviewed basic fault-tolerant methods in non-real-time and real-time systems in this chapter. Redundancy in the form of duplicated or extra hardware or software is a fundamental technique for fault tolerance. The basic types of redundancy are hardware, information, time, and software redundancies. Hardware redundancy is commonly realized by duplicating hardware components and running the software in all of the components and then voting to obtain the correct result. Information redundancy is achieved mainly by coding using extra data to check errors. Time redundancy is accomplished by repeating the computation in different time points and then comparing the results. Software redundancy can be employed as single-version or multiple-version techniques. Extra software modules such as exception handling and checkpointing are added to a single software module to detect faults and recover from them in the former and recovery blocks, N-version programming, and N-self-checking programming methods are employed in the latter. Multiple-version software redundancy methods commonly use N versions of the software, run them, and compare the results obtained.

All of these redundancy methods can and are implemented in real-time systems. Additionally, the main requirement from a real-time system which is the meeting of the deadlines of tasks must be provided in the presence of faults. We provided a survey of some research studies on this topic based on static and dynamic scheduling. A detailed review of the some of the topics we have discussed in this chapter can be found in [8], and fault-tolerant design methods are reviewed in [5].

Table 12.2 An example four 4-bit data for Exercise 1

c_4	c_3	c_2	c_1	c_0
0	1	1	0	
1	0	1	1	
1	0	1	0	
1	1	1	1	

12.10 Exercises

1. Given the four 4-bit binary data in Table 12.2, work out the horizontal and vertical odd parity bits.
2. Work out the 4-bit check word to be appended to the binary data 1000 0110 11 using CRC error detection method using the polynomial word 11011.
3. Compare reliable multicast ordering protocols FIFO, casual, and atomic.
4. Implement the FIFO multicast protocol using the structure of DRTK. Write the C code with brief comments.
5. Propose a token-based protocol for total ordered multicast. Start design by the FSM of the protocol. Write the C code that will interface to DRTK with brief comments.

References

1. Burns A, Wellings A (2009) Real-time systems and programming languages, 4th edn, Chap. 2. Addison-Wesley
2. Chandra TD, Toueg S (1996) Unreliable failure detectors for reliable distributed systems. J ACM 43(2):225–267
3. Chen L, Avizienis A (1978) N-version programming: a fault tolerance approach to reliability of software operation. In: Digest of 8th annual international symposium on fault tolerant computing, pp 3–9
4. Defago X, Schiper A, Urban P (2004) Totally ordered broadcast and multicast algorithms: taxonomy and survey. J ACM Comput Surv 36(4):372–421
5. Dubrova E (2013) Fault tolerant design, Chap. 4. Springer, New York
6. Lee PA, Anderson T (1990) Fault tolerance: principles and practice, 2nd edn. Springer
7. Kandasamy N, Hayes JP, Murray BT (2000) Task scheduling algorithms for fault tolerance in real-time embedded systems. In: Avresky DR (ed) Dependable network computing. The Springer international series in engineering and computer science, vol 538. Springer, Boston, MA
8. Koren I, Krishna CM (2007) Fault tolerant systems. Morgan-Kaufman
9. Lamport L (1978) Time, clocks, and the ordering of events in a distributed system. Commun ACM 21(7):558–565
10. Laprie JC (1985) Dependable computing and fault tolerance: concepts and terminology. In: The 15th international symposium on fault-tolerant computing, pp 2–11

11. Oh Y, Son S (1994) Scheduling hard real-time tasks with tolerance of multiple processor failures. Microprocess Microprogr 40:193–206
12. Punnekkat S (1997) Schedulability analysis for fault tolerant real-time systems. PhD thesis, Department of Computer Science, University of York
13. Tanenbaum AS, Van Steen M (2006) Distributed systems, principles and paradigms, Chap. 8. Prentice Hall

Case Study: Environment Monitoring by a Wireless Sensor Network

13

13.1 Introduction

A wireless sensor network (WSN) consists of small computational nodes each equipped with wireless communication facilities. WSNs are commonly used in hazardous environments for monitoring and rescue operations. We will show the implementation of a case study: monitoring of a complex building by a wireless sensor network using the sample distributed real-time operating system kernel DRTK. The WSN will be used for monitoring a building and protection against fire, floods, and intruders. This small project will help to gain more insight into many concepts reviewed throughout the book. The system designed will be a distributed embedded real-time system and we will implement all of the design steps described until now, starting with the high-level design up to and including the C language code. We start with the requirement specifications and form the SRS document in the usual manner. Timing analysis is done as the next step and the high-level design considering the hardware–software co-design is then performed. The tool employed for high-level design is data flow diagrams (DFDs), and we start this procedure by forming the context diagram of the system first. The detailed design of the system involves designing C functions to be realized by the tasks of DRTK.

We provide an alternative implementation of this project without using DRTK. This time, the steps to be followed are the ones described as the practical design method in Chap. 10 and the functions are realized by POSIX threads with all the POSIX thread synchronization methods becoming available. For thread communication, we use the novel inter-thread communication method described in Chap. 4.

© Springer Nature Switzerland AG 2019

K. Erciyes, *Distributed Real-Time Systems*, Computer Communications and Networks, https://doi.org/10.1007/978-3-030-22570-4_13

13.2   Design Considerations

We will use this case study to implement all of the steps of distributed real-time system design specified as follows.

1. *Requirement Specifications*: To state the user/customer requirements formally. We will produce an SRS document at the end of this phase.
2. *Timing Analysis*: We need to check how timing requirements are to be met before performing high-level design.
3. *High-level Design*: We will be using data flow diagrams to gradually refine system functions to task level.
4. *Detailed Design*: We will implement the required functions using real-time tasks that will be represented by tasks of DRTK.
5. *Coding*: All of the codings will be done in C programming language.

We will describe two ways of performing the design, using DRTK as the operating system kernel and using the POSIX interface as stated.

13.3   Requirement Specifications

The main requirement is the monitoring of a large, complex building by a WSN against hazardous conditions. The SRS document specifies the following:

1. **Introduction**: The building to be monitored has 25 floors each with 20–30 rooms and a basement and car park facilities. The building is to be protected against fire, floods, and water leakages. The monitoring of temperature, smoke, and humidity should be performed continuously and any intruder should be detected.
2. **Functional Requirements**

   a. The temperature of every room, every floor, and various other places in the building should be monitored. There will be LED displays for temperature values in every room, every floor, and various other places in the building. When the temperature of a location is over or below a specified limit value, sound and display alarms should be activated. The management computer should detect the location where the values are beyond limits.
   b. Protection against fire is to be provided by detecting smoke in every room, every floor, and various other places in the building. There will be sound and light alarms in every room, every floor, and various other places in the building. In case of fire, the sprinters in the corridors will be switched on and the sound and light alarms will be activated. The management computer should detect the location of the fire floor.
   c. The humidity of the building should be monitored to take actions against floods or other water leakages. There will be sound and light alarms for

humidity in every room, every floor, and various other places in the building. When there is high humidity, the humidity alarm should be set and the manager should be informed of the location.

d. When the building is closed, any intruder with its location should be detected, intruder alarm should be set, and the manager computer should be informed about the location of the intruder.

e. All of the values of these parameters in the building should be displayed in the managerial computer in condensed form such as average temperature and humidity values of each floor.

3. **Nonfunctional Requirements**

a. The system should be fault tolerance. Failing to detect smoke, heat, or humidity in one location should not cause a total system failure.

b. The system should be guaranteed to work for 5 years.

This document is a contract between the user/customer and the system designer. Moreover, it will guide us throughout the design process.

13.4 Timing Analysis and Functional Specifications

A closer look at requirements shows that timing constraints in the system to be designed are not tight. We need to input slowly changing temperature and humidity values into the system periodically, and the intruder detection may happen at any time. Therefore, the system has both time-triggered and event-triggered characteristics. Employing WSN nodes each of which communicates using multi-hop communications is considered convenient for the design since cabling is eliminated and the processing to be done by each sensor node is not substantial.

This is the point we need to consider the hardware–software co-design. Employing a WSN that uses multi-hop communication affects the software structure of the system. A spanning tree rooted at the *sink* node of the WSN, the managerial computer in this case, is commonly used for multi-hop communications between the sink and the ordinary nodes in a WSN. We therefore need to construct a spanning tree of the WSN initially. The required condensed form of display by the manager means that some form of preprocessing is to be performed at intermediate nodes before sensor data is delivered to the sink. This requirement can be met by grouping the sensor nodes that are physically close to each other in *clusters* with a *clusterhead* to manage data operations and the communication traffic. Note that considering special hardware initially affected the design of the high-level software significantly.

13.5 Spanning Tree and Clustering

A spanning tree T of a graph $G = (V, E)$ is an acyclic subset of G with the same set of vertices, that is, $T = (V, E')$ where $E' \subseteq E$. In other words, each vertex is part of the spanning tree but only a subset (or equal set) of edges belong to T. Forming a spanning tree of a WSN is a convenient method of communication in these networks since senders and receivers of a message can be selectively defined. Broadcasting from the sink node can be achieved simply by sending the message to children by each parent node except the leaf nodes. After construction, the WSN nodes become part of a spanning tree and are partitioned into a number of clusters.

Having a spanning tree and a cluster structure can be obtained by the algorithm described in [1]. The message types in this algorithm are PROBE, ACK, and NACK, and each node in the network becomes either a clusterhead (CH), intermediate node (INODE), or a leaf (LEAF) of the spanning tree at the end of the algorithm. The *sink* node of the WSN starts the algorithm by broadcasting a PROBE message to all neighbors within its transmission range. Any node that receives the message for the first time marks the sender as its parent by sending an ACK message to the sender and rejecting any other PROBE messages received thereafter, by the NACK message. This procedure provides forming the spanning tree; however, we need another mechanism to form the clusters. This is achieved by reserving a *hop_count* field in the PROBE messages transmitted. Any node that receives a PROBE message for the first time also checks the *hop_count* in the message and if this value is equal to a specified value of cluster depth (*clust_depth*), it assigns itself to CH state and resets *hop_count* before transmitting the PROBE message to neighbors. Any other node that receives a *hop_count* less than the specified value changes its state to INODE and increments *hop_count* in the message before transmitting it to neighbors. The FSM of this algorithm is depicted in Fig. 13.1.

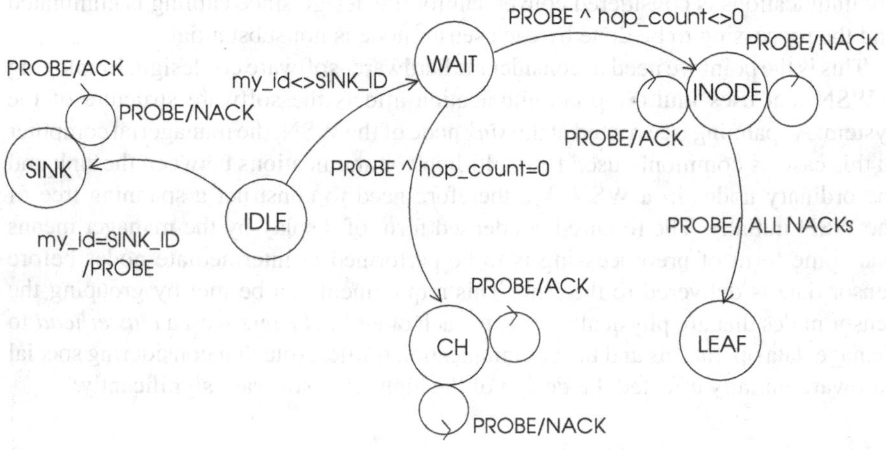

Fig. 13.1 FSM of the *Cluster* task

Table 13.1 FSM table for the cluster task

	Inputs	PROBE	ACK	NACK
States	IDLE	NA	NA	NA
	WAIT	*act*_10	NA	NA
	INODE	*act*_20	*act*_21	*act*_22
	LEAF	*act*_30	NA	NA

Starting from IDLE state, a test on the node identifier classifies the node as the sink or any other node. Note that this state is internal without waiting for any input. A node at INODE state may be transferred to LEAF state if all of its neighbors have responded with NACK messages to its PROBE messages meaning it may not have any children, and therefore it is a leaf node. The FSM table that will guide us in implementing the algorithm is shown in Table 13.1.

We will assume that each node has a unique identifier while implementing the code. We will further assume that each node of the WSN is equipped with the sample operating system kernel DRTK, allowing the use of all DRTK system calls we have developed along with network communication methods and task management routines. One way of realizing the FSM is shown in the C code below. We have a header file "tree_clust.h" which defines message types and states of nodes shown below.

```
// tree_clust.h

#define     PROBE       1
#define     ACK         2
#define     NACK        3
#define     CH          1
#define     INODE       2
#define     LEAF        3
#define     N_STATE     4
#define     N_INPUT     3
```

We can now write the code for the algorithm called *Tree_Cluster* based on the FSM table actions. Note that the first entry of the data field of the message is reserved for the *hop_count* variable.

```
/*****************************************************************
                    Clustering Procedure
*****************************************************************/
/* tree_clust.c */ #include "tree_clust.h"

fsm_table_t sender_FSM[N_STATE][N_INPUT];
  mailbox_ptr_t mbox_pt;
  data_unit_ptr_t data_pt;
  ushort my_id, my_parent;
  int my_state;
```

```
void act_10() {
    my_parent=data_pt->MAC_header.sender_id;
    my_neighbors[neigh_index++]=my_parent;
    if (data_pt->data[0]==N_HOPS+1) {
      current_state=CH;
      data_pt->data[0]=0;
    }
    else current_state=INODE;
    data_pt->MAC_header.sender_id=my_id;
    data_pt->MAC_header.type=ACK;
    data_pt->data->[0]++;
    send_mailbox_notwait(&(task_tab[System_Tab.DL_Out_id])
            ->mailbox_id,data_pt);
}

void act_20() {
    data_pt->sender=my_id;
    data_pt->MAC_header.type=NACK;
    send_mailbox_notwait(&(task_tab[System_Tab.DL_Out_id])
            ->mailbox_id,data_pt);
}

void act_21() {
    my_neighbors[neigh_index++]=data_pt->sender;
    return_data_unit(System_Tab.userpool1,data_pt);
}

void act_22(){
    n_rejects++;
    if(n_rejects==N_NEIGHBORS)
      my_state=LEAF;
    return_data_unit(System_Tab.userpool1,data_pt);
}

TASK Tree_Cluster() {

  clust_FSM[0][0]=NULL;    clust_FSM[0][1]=NULL;
  clust_FSM[0][2]=NULL;    clust_FSM[1][0]=act_10;
  clust_FSM[1][1]=NULL;    clust_FSM[1][2]=NULL;
  clust_FSM[2][0]=act_20; clust_FSM[2][1]=act_21;
  clust_FSM[2][2]=act_22; clust_FSM[3][0]=act_20;
  clust_FSM[3][1]=NULL;    clust_FSM[3][2]=NULL;

  mboxpt=&(tcb_tab[current_pid].rec_mailbox);
  current_state=IDLE;

  while(TRUE) {
   data_pt=recv_mailbox_wait((task_tab[current_tid]).mailbox_id);
   (*sender_FSM[current_state][data_pt->type])();
  }
}
```

A WSN partitioned into five clusters $C_1, \ldots, C_6$ with $n\_hops = 2$ over a spanning tree is shown in Fig. 13.2. Note that periodically invoking the algorithm by the *sink* allows building clusters over a spanning tree in a mobile ad hoc network by this algorithm.

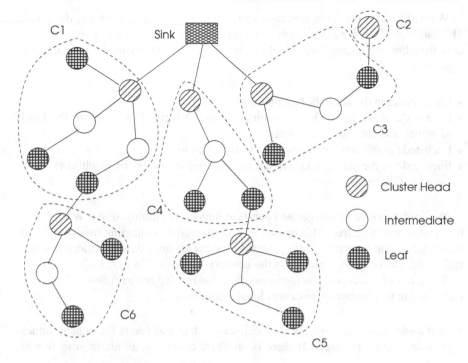

Fig. 13.2 A clustered WSN over a spanning tree

13.6 Design Considerations

We will have the high-level and detailed design of the WSN monitoring system based on the type of the nodes. A node can be one of four types: a *leaf* node, an *intermediate* node, a *clusterhead*, and the *sink*. We need to design the DFDs for each of these based on the requirements in the following sections. The types of messages traversing the edges of the spanning tree will be of following types.

- *Node message*: Transferred from a leaf node of a cluster to its parent. It contains the three sensor values and alarm conditions on these sensors. This message type is also used for local sensor information.
- *Intermediate message*: Transferred from an intermediate node to another intermediate node or to a clusterhead. This message contains the sensor values of the children of the intermediate node and alarm conditions on these sensors. Note that this node does not process non-local sensor values; it only merges these values.
- *Clusterhead message*: This message may be transferred between any type of node to be transferred to the sink. It carries the average sensor values at the cluster and possible alarm conditions.

We will be forming these message structures as the node software is designed. The data area of the data unit needs to be formed as a union of these structures. We have the following assumptions which can be realized by modifying the *Tree_Cluster* algorithm.

- Each cluster in the network has a unique identifier.
- Each node in the network has a unique identifier formed by merging the local identifier with the cluster identifier.
- Each node in the network knows its cluster identifier.
- Each node in the network knows the number and identifiers of its children from the *Tree_Cluster* algorithm.

We need the basic convergecast operation frequently employed in a WSN structured into a spanning tree. This operation is performed by collecting messages from the children, and then transferring them to the parent until data is collected at the sink node. The message structure in the network is depicted in Fig. 13.3.

The types of nodes and the messages they can receive are as follows. Note that each node in this hierarchy processes local sensor data.

- *Leaf node*: This node obtains its local sensor data and sends the sensor values in *node_msg* to its parent. If there is an alarm condition, an alarm_msg is sent

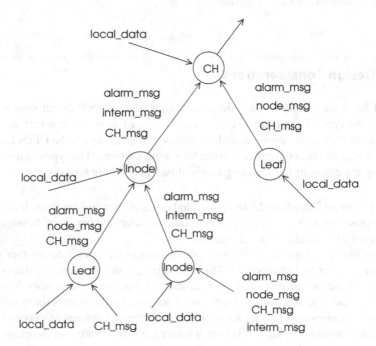

Fig. 13.3 Message hierarchy in the network

immediately to the parent. It also sends and incoming clusterhead message (*CH_msg*) from lower clusters to its parent.

- *Intermediate node*: The messages to intermediate node (INODE) are the *alarm_msg*, *node_msg* from the leaf, *interm_msg* from any child that is not a leaf and any incoming *CH_msg* from lower level clusters.
- *Clusterhead*: The messages to this node are similar to the input messages to the intermediate node.

13.7 Leaf Node

The leaf node is the leaf of the cluster which may not be the leaf of the spanning tree. Hence, it may receive messages from lower layer clusterheads. However, due to the structure of the spanning tree, it may not receive messages from an intermediate node. We will first describe the high-level design of this node, continuing with the implementation.

13.7.1 High-Level Design

A leaf node basically collects temperature and humidity sensor data periodically and sends this data to its parent. Upon receiving an interrupt caused by an intruder, it also sends this data to the parent and sets the alarm button on using the following steps.

1. Collect sensor data periodically and send it to parent.
2. If data is out of range, set data alarm on.
3. When an intruder is detected, send this data to parent immediately and set the intruder alarm on.
4. Upon reception of a clusterhead message from a lower level clusterhead, send it to parent.

Therefore, the main inputs to a leaf node are the sensor inputs and the cluster-head message as depicted in the context diagram of Fig. 13.4. The outputs are the intermediate message and alarm displays.

The level 1 diagram based on the context diagram can be formed as in Fig. 13.5. The data from the sensors is received periodically by the *Temp* and *Humid* tasks which first check whether this data is in the allowed range and deposit the data in the *alarm_data* data store to awake the *Alarm* task to display when data is out of range. Data is stored in *sensor_data* store in any case to be sent to the parent. Any intruder in the monitored region is detected by the *Accel* task which stores data in both stores to activate alarm and report to the parent. The *Leaf_In* task continuously waits for messages from the network and sends any incoming clusterhead message to its parent by the *Leaf_Out* task.

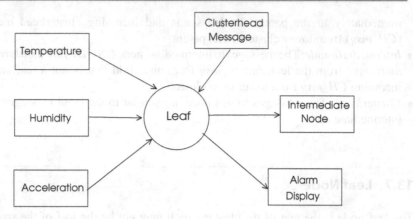

Fig. 13.4 Context diagram of the leaf node

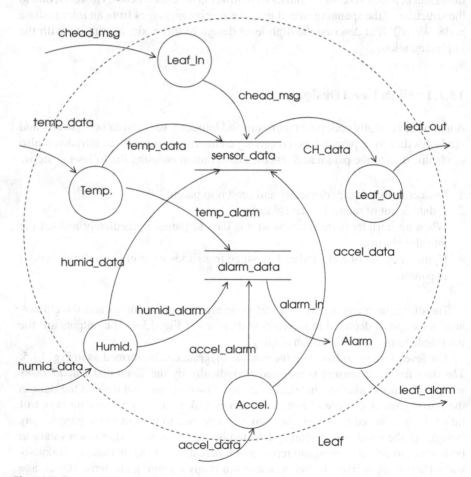

Fig. 13.5 Level 1 DFD of the leaf node

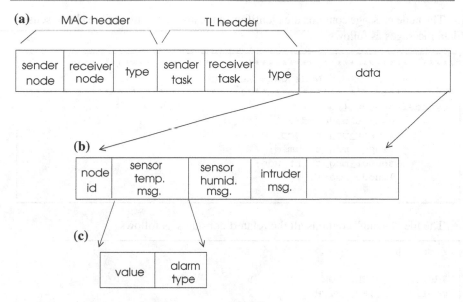

Fig. 13.6 **a** General message format, **b** node message, **c** sensor message

13.7.2   Detailed Design and Implementation

First thing to note is that the bubbles in Level 1 diagram can be conveniently represented by procedures. We arrive at this conclusion by considering what each task must do, using intuition. For example, the *Temp* task needs to delay itself for a period of time, then wake up, read the temperature value, check its range, and store the data in data stores. The general format of the message that is passed between tasks and traversed in the network is specified in Fig. 13.6a, and the parsing of data area for the node message is depicted in Fig. 13.6b. The sensor message shown in Fig. 13.6c consists of sensor value and alarm condition as the lowest layer data message. The MAC header shows the sending and receiving nodes and the type at this level such as *unicast*, *multicast*, or *broadcast* message or a control message such as acknowledgment or negative acknowledgment. The transport layer header is used for inter-task communications since main function of this layer is to provide communication between user/application tasks.

The message structure that is filled by the sensor input tasks has the following structure with the sensor value and the alarm condition.

```
/******************************************************************
                    Sensor Message Structure
******************************************************************/
typedef struct {
        double value;
        int alarm_t;
        } sensor_msg_t, alarm_msg_t, *sensor_msg_ptr_t;
```

The node message contains the identifier of the node and three of the above sensor data messages as follows.

```
/*****************************************************************
                      Node Data Message
*****************************************************************/
typedef struct {
        ushort node_id;
        sensor_msg_t temp;
        sensor_msg_t humid;
        sensor_msg_t intrude;
        } node_msg_t;
```

The file "local.h" contains all the related constants as follows.

```
//local.h

#define     TEMP_LOW          5
#define     TEMP_HIGH         2
#define     HUMID_LOW        12
#define     HUMID_HIGH      120
#define     ERR_TEMP_LOW     -1
#define     ERR_TEMP_HIGH    -2
#define     ERR_HUMID_LOW    -3
#define     ERR_HUMID_HIGH   -4
#define     ERR_INTRUDER     -5
#define     N_CLUS_NODE      20
#define     N_CHILD_MAX      10
#define     SENSOR_MSG        1
#define     NODE_MSG          2
#define     INTERM_MSG        3
#define     CHEAD_MSG         4
#define     TEMP_MSG          1
#define     HUMID_MSG         2
#define     INTRUDER_MSG      3
```

Normally, we would have used FSM model to represent the bubbles in the Level 1 diagram but there are not many external interactions of the functions at this level, and hence we can proceed without FSMs. We assume DRTK is present at each node of the WSN which allows us to use mailboxes as data storage places, and devices can be attached to sensors. The first task we will code is *Temp* which delays itself and reads sensor data from the device control block of the temperature device when awake. After checking limits, it sends data to the related mailboxes as shown below. We will assume task identifiers, and the device identifiers are declared in the system table. A sensor input task needs to fill sensor message with the sensor data before sending the message to the *Leaf_Out* task as shown in the code below.

```
/*****************************************************************
                        Temperature Task
*****************************************************************/

TASK Temp() {
 data_unit_ptr_t data_pt1, data_pt2;
 int dev_id, mbox_id1, mbox_id2,mbox_id3;
 ushort dev_id=System_tab.temp_devid;
 ushort mbox_id1=task_tab[current_tid].mailbox_id;
 ushort mbox_id2=task_tab[System_Tab.Leaf_Out_id].mailbox_id;
 ushort mbox_id3=task_tab[System_Tab.Alarm_id].mailbox_id;

 while(TRUE) {
   delay_task(current_tid, System_Tab.LOCAL_DELAY);
   data_pt1=get_data_unit(System_Tab.userpool1);
   read_device(dev_id,&(data_pt1->data)->value,sizeof(double));
   data_pt1->TL_header.type=TEMP_MSG;
   if (data_pt1->data.value < TEMP_LOW ||
            data_pt1->data.value > TEMP_HIGH) {
     data_pt2=get_data_unit(System_Tab.userpool1);
     data_pt1->TL_header.type=ALARM_MSG;
     if (data_pt1->data.value < TEMP_LOW) {
       data_pt1->data.sensor_msg.alarm_t=ERR_TEMP_LOW;
       data_pt2->data.alarm_msg.alarm_t=ERR_TEMP_LOW;}
     else if (data_pt1->data.value > TEMP_HIGH) {
       data_pt1->data.sensor_msg.alarm_t=ERR_TEMP_HIGH;
       data_pt2->data.alarm_msg.alarm_t=ERR_TEMP_HIGH;
     send_mailbox_notwait(mbox_id3,data_pt2); }
   }
   send_mailbox_notwait(mbox_id2,data_pt1);
 }
}
```

The *Humid* task which records humidity data works similarly except it tests humidity allowed levels as shown in the code below.

```
/*****************************************************************
                        Humidity Task
*****************************************************************/

TASK Humid() {
 data_unit_ptr_t data_pt1, data_pt2;
 int dev_id, mbox_id1, mbox_id2,mbox_id3;
 dev_id=System_tab.humid_devid;
 ushort mbox_id1=task_tab[current_tid].mailbox_id;
 ushort mbox_id2=task_tab[System_Tab.Leaf_Out_id].mailbox_id;
 ushort mbox_id3=task_tab[System_Tab.Alarm_id].mailbox_id;

  while(TRUE) {
```

```
    delay_task(current_tid, System_Tab.LOCAL_DELAY);
    data_pt1=get_data_unit(System_Tab.userpool1);
    read_device(dev_id,&(data_pt1->data),sizeof(double));
    data_pt1->TL_header.type=HUMID_MSG;
    if (data_pt1->data.value < HUMID_LOW ||
        data_pt1->data.value > HUMID_HIGH) {
     data_pt2=get_data_unit(System_Tab.userpool1);
     data_pt1->TL_header.type=ALARM_MSG;
     if (data_pt1->data.value  < HUMID_LOW) {
        data_pt1->data.sensor_msg.alarm_t=ERR_HUMID_LOW;
        data_pt2->data.alarm_msg.alarm_t=ERR_HUMID_LOW; }
     else if (data_pt1->data.value > HUMID_HIGH) {
       data_pt1->data.sensor_msg.alarm_t=ERR_HUMID_HIGH;
       data_pt2->data.alarm_msg.alarm=ERR_HUMID_HIGH; }
     send_mailbox_notwait(mbox_id3,data_pt2);
     }
    send_mailbox_notwait(mbox_id2,data_pt1);
  }
}
```

The *Accel_Int* task is event-driven and is awaken by an interrupt from the acceleration sensor. This task always waits for an interrupt and sends an alarm message to the alarm task when awaken.

```
/****************************************************************
                Acceleration High Interrupt Handler
****************************************************************/
TASK Accel_Int() {
  data_unit_ptr_t data_pt;

 ushort mbox_id1=task_tab[current_tid].mailbox_id;
 ushort mbox_id2=task_tab[System_Tab.Leaf_Out_id].mailbox_id;
 ushort mbox_id3=task_tab[System_Tab.Alarm_id].mailbox_id;

 while(TRUE) {
   data_pt=recv_mailbox_wait(mbox_id1);
   data_pt->TL_header.type=ALARM_MSG;
   data_pt->data.alarm_t=ERR_INTRUDER;
   send_mailbox_notwait(mbox_id2,data_pt);
   send_mailbox_notwait(mbox_id3,data_pt);
 }
}
```

The job of *Leaf_Alarm* task is simply to wait for any alarm messages in its mailbox, decode them, and activate associated displays. We will assume there is a single driver to activate all alarms which checks the type field in the *write_dev* function.

```
/ *************************************************************
                        Alarm Task
  ************************************************************* /
TASK Leaf_Alarm(ushort dev_id) {

    data_unit_ptr_t data_pt;
    int err_type;
    ushort mbox_id=task_tab[current_tid].mailbox_id;
    ushort dev_id=System_tab.alarm_devid;

    while(TRUE) {
      data_pt=recv_mailbox_wait(mbox_id);
      err_type=data_pt->data.sensor_msg.alarm_t;
      write_dev(dev_id,&err_type,sizeof(int));
    }
}
```

The *Leaf_In* task waits for messages from the network in its mailbox and transfers these to the *Leaf_Out* task as shown in the code below. We assume the existence of a data link task which receives messages using network drivers and deposits them in the mailbox of the *Leaf_In* task. For a specific network, the function of this task may be embedded in the data link task.

```
/ *************************************************************
                        Leaf In Task
  ************************************************************* /
TASK Leaf_In() {

    data_unit_ptr_t data_pt;
    ushort mbox_id=task_tab[current_tid].mailbox_id;
    ushort dev_id=System_Tab.alarm_dcbid;

    while(TRUE) {
      data_pt=recv_mailbox_wait(mbox_id);
      if (data_pt->TL_header.type==CHEAD_MSG) {
        mbox_id=task_tab[System_Tab.Leaf_Out_id].mailbox_id;
        send_mailbox_notwait(mbox_id,data_pt);
      }
    }
}
```

The final task to be implemented is *Leaf_Out* task which waits for periodic sensor messages, collects them into a single node message, and transfers them to its parent.

It calls the function *check_local* to find the type of message and do necessary house-keeping shown below. This function also transfers any incoming CH message and alarm message without changing their contents immediately as shown in the code below.

```
data_unit_ptr_t data_pt, data_pt2;
ushort my_id, parent_id, sensor_count=0;
ushort mbox_id1=&(task_tab[current_tid])->mailbox_id;
ushort mbox_id2=&(task_tab[System_Tab.DL_Out_id])->mailbox_id;

int check_local(ushort type, data_ptr_t data_pt) {

    switch(type) {
       case TEMP_MSG:
          data_pt->data.node_msg.temp_val=recvd_pt
                  ->data.sensor_msg.value;
          sensor_count++; break;
       case HUMID_MSG:
          data_pt->data.node_msg.humid_val=recvd_pt
                  ->data.sensor_msg.value;
          sensor_count++; break;
       case ALARM_MSG:
       case CHEAD_MSG:
          data_pt2->MAC_header.sender_id=System_Tab.this_node;
          data_pt2->MAC_header.receiver_id=my_parent;
          send_mailbox_notwait(mbox_id2,data_pt2);
          data_pt2=get_data_unit(System_Tab.userpool1);
          break;
       default: return(NOT_FOUND);
    }
  return (DONE);
}
```

The *Leaf_Out* task, code of which is shown below, calls the *check_local* function and when two sensor values are received by *sensor_msg* messages, it puts them into a single *node_msg* to be transferred to the parent.

```
/****************************************************************
                    Leaf Out Task
****************************************************************/
TASK Leaf_Out() {

    data_pt=get_data_unit(System_Tab.userpool1);
    data_pt2=get_data_unit(System_Tab.userpool1);
    while(TRUE) {
       delay_task(current_tid, System_Tab.LOCAL_DELAY);
       recvd_pt=recv_mailbox_wait(mbox_id1);
       check_local(recvd_pt->TL_header.type, data_pt);
       if (sensor_count==2) {
          data_pt->TL_header.type=NODE_MSG;
          data_pt->data.node_msg.node_id=System_Tab.this_node;
```

```
        data_pt->MAC_header.sender_id=System_Tab.this_node;
        data_pt->MAC_header.receiver_id=my_parent;
        send_mailbox_notwait(mbox_id2,data_pt);
        data_pt=get_data_unit(System_Tab.userpool1);
        sensor_count=0;
        data_pt=get_data_unit(System_Tab.userpool1);
    }
  }
}
```

13.8   Intermediate Node

An intermediate node is any node other than the leaves and the CHs. The main function of this type of node is to collect all data from its children on the spanning tree formed, merge this data with its local data, and send all of this data upstream to its parent. Possible messages to this type of node are the leaf, intermediate, and CH messages.

13.8.1   High-Level Design

The context diagram of the intermediate node is depicted in Fig. 13.7 with all the external entities shown. This type of node collects its own environment data, merges this data into a single message, and sends it to another intermediate node or to its clusterhead.

Level 1 DFD of the Intermediate Node is depicted in Fig. 13.8. The local environment monitoring of this node will be the same as the leaf, so we will not repeat the functions of the *Temp*, *Humid*, and *Accel* tasks and data structures. We have the task *IMerge* which basically merges all received temperature and humidity values of all children, notes alarm conditions in received values, and activates alarms if needed by depositing alarm messages (*comp_al*) in the *comp_alarm* data storage.

We begin by specifying the message contents that is transferred between an intermediate node and another intermediate node/cluster head. This message contains the local data of all children of the intermediate node as shown in Fig. 13.9.

The code for this structure can be formed as follows.

```
/*****************************************************************
                 Intermediate Message Structure
 *****************************************************************/

 typedef struct {
         ushort inode_id;
         ushort n_msg;
         node_msg_t node_msg[N_CHILD_MAX];
         } interm_msg_t, *interm_msg_ptr_t;
```

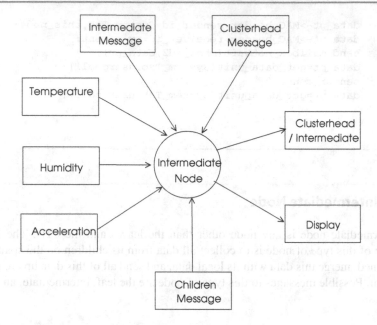

Fig. 13.7 Context diagram of the intermediate node

13.8.2   Detailed Design and Implementation

We need a front-end task called *Inter_In* which collects data from all of the children of the node and merges this data with the local data to be sent upstream to another intermediate node or to the clusterhead. This task also may receive cluster head message which is forwarded to the parent. We will assume all of the nodes are time synchronized and task delays are close to each other so that data from children arrive at this task in a small time frame, and all nodes function correctly without failure.

```
/*****************************************************************
                        Inter_In Task
*****************************************************************/
TASK Inter_In() {

    data_unit_ptr_t data_pt;
    ushort mbox_id1=task_tab[current_tid].mailbox_id;
    ushort mbox_id2;

    while(TRUE) {
     delay_task(current_tid,System_Tab.LOCAL_DELAY);
     data_pt=recv_mailbox_wait(mbox_id1);
     switch(data_pt->TL_header.type) {
      case LEAF_MSG    :
      case INTERM_MSG  :
        mbox_id2=&(task_tab[System_Tab.Imerge_id])->mailbox_id;
```

```
        break;
     case CHEAD_MSG    :
     case ALARM_MSG    :
        mbox_id2=&(task_tab[System_Tab.Inter_Out_Id])->mailbox_id;
     }
     send_mailbox_notwait(mbox_id2,data_pt);
   }
}
```

The intermediate node merging task (*Imerge*) inputs all of the data pertaining to a node from the *Inter_In* task and merges them in a single message. When all the children of the intermediate node have sent their data, the single message may be sent to the output task *Inter_Out* to be transferred to the parent. Note that inputs to this task are the local node data and any intermediate child data which need to be merged into an intermediate message. The variable *child_count* is used to count the number

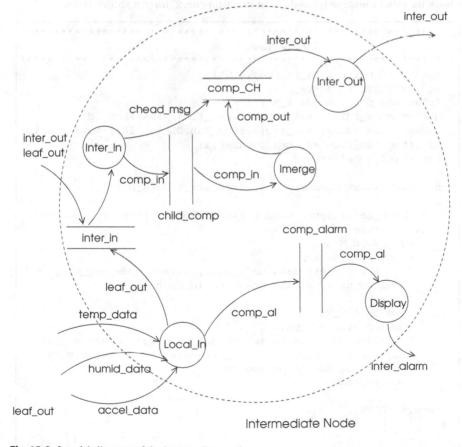

Intermediate Node

Fig. 13.8 Level 1 diagram of the intermediate node

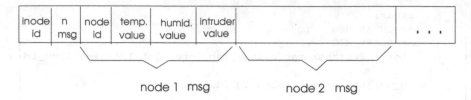

Fig. 13.9 Intermediate message structure

of messages received and when this value is equal to the number of the children of the intermediate node, the message can be transferred to the parent. We assume the local sensor values are combined into a node message and delivered to this task by the *Local_In* task which is not shown. This task may receive local sensor messages, and needs to put them into a *node_msg*, or it may receive alarm or CH messages which need to be transferred immediately. It uses the *check_local* function of the leaf module for this purpose and if the message type is not recognized in this function, check for other message types done in function *check_interm* shown below.

```
/*****************************************************************
                       IMerge Task
*****************************************************************/
  data_unit_ptr_t data_pt, recvd_pt;
  node_msg_ptr_t nddata_pt;
  inter_msg_ptr_t intdata_pt;
  ushort mbox_id1=&(task_tab[current_tid])->mailbox_id;
  ushort mbox_id2=&(task_tab[System_Tab.DL_Out_id])->mailbox_id;
  ushort child_count=0, msg_count=0,i;
  int check_local(int);

 int check_interm(ushort type, data_ptr_t data_pt) {

   if(check_local(type, data_pt)==NOT_FOUND || sensor_count==0)
    switch(type) {
      case NODE_MSG :
        memcpy(nddata_pt,
            &(recvd_pt->data.node_msg), sizeof(node_msg_t));
        msg_count++; nddata_pt++; child_count++;
        break;
      case INTERM_MSG:
        memcpy(intdata_pt,
            &(recvd_pt->data.interm_msg), sizeof(interm_msg_t));
        msg_count=msg_count+recvd_pt->data.interm_msg.n_msg;
        nddata_pt=nddata_pt+msg_count;
        child_count++;
    }
  return (DONE);
}
```

The *Imerge* task calls this function to find the type of message and do the necessary data transfers based on the type. When it has messages from all of its children, the convergecast data may be sent as a single *intermediate* message to the parent.

```
TASK Imerge() {

    data_pt=get_data_unit(System_Tab.userpool1);
    nddata_pt=&(data_pt->data.node_msg);
    intdata_pt=&(data_pt->data.interm_msg);
    while(TRUE) {
      delay_task(current_tid,System_Tab.LOCAL_DELAY);
      recvd_pt=recv_mailbox_wait(mbox_id1);
      check_interm(recvd_pt->TL_header.type,data_pt);
      if (sensor_count==2) {
        memcpy(nddata_pt,
            recvd_pt->data.node_msg, sizeof(node_msg_t));
        msg_count++; nddata_pt++; child_count++;
        sensor_count=0;
      }
      if(child_count==&(System_Tab.n_children+1)) {
        data_pt->data.interm_msg.inode_id=System_Tab.this_node;
        data_pt->TL_header.type=INTERM_MSG;
        data_pt->TL_header.sender_id=System_Tab.this_node;
        data_pt->data.interm_msg.n_msg=msg_count;
        data_pt->MAC_header.sender_id=System_Tab.this_node;
        data_pt->MAC_header.receiver_id=my_parent;
        send_mailbox_notwait(mbox_id2,data_pt);
        data_pt=get_data_unit(System_Tab.userpool1);
        nddata_pt=&(data_pt->data.node_msg);
        intdata_pt=&(data_pt->data.interm_msg);
        child_count=msg_count=0;
        data_pt=get_data_unit(System_Tab.userpool1);
      }
    }
}
```

The code for Alarm task is omitted as it will be very similar to the code of the alarm task of the leaf. The output task *Inter_Out* simply receives the message in its mailbox and deposits the message in the mailbox of the data link layer out task. Note that we could manage with the *Imerge* task performing this operation as well as merging; however, having tasks dedicated to input and output operations is convenient in general so as to modify and enhance input/output functions when needed.

```
/***********************************************************
                     Inter_Out Task
***********************************************************/

TASK Inter_Out() {

    data_unit_ptr_t data_pt;
    ushort mbox_id=task_tab[current_tid].mailbox_id, mbox_id2;
    mbox_id2=task_tab[System_Tab.DL_Out_Id].mailbox_id
    while(TRUE) {
      data_pt=recv_mailbox_wait(mbox_id);
      data_pt->MAC_header.sender_id=System_Tab.this_node;
      data_pt->MAC_header.receiver_id=my_parent;
      send_mailbox_notwait(mbox_id,data_pt);
    }
}
```

13.9   Clusterhead

The clusterhead node performs local sensing as all other nodes other than the sink do. It also performs *data aggregation* by computing the average values of the temperature values and humidity values and forming a message structure shown in Fig. 13.10. This *clusterhead message* has a cluster field containing the cluster identifier, the average temperature and humidity values for the cluster, and node identifiers and the alarm conditions in the cluster if they exist. The alarm condition for each node in a cluster should be delivered to the sink node, and hence we have such fields to specify this condition.

data area of the message

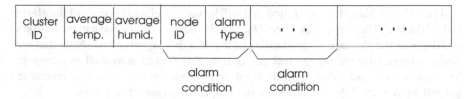

Fig. 13.10 Clusterhead message data area structure

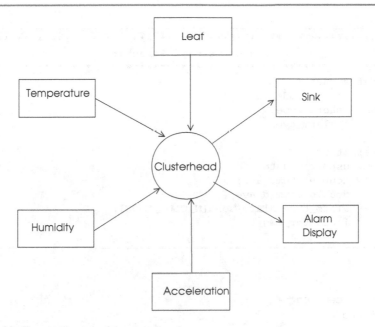

Fig. 13.11 Context diagram of the clusterhead

13.9.1 High-Level Design

Level 0 DFD of the clusterhead is depicted in Fig. 13.11. We may have leaf, inter-mediate node, and clusterhead type of messages coming to this node. Its output may go to a leaf node, clusterhead, or to the sink.

The level 1 DFD diagram in Fig. 13.12 displays the task structure of this node which is similar to the level 1 DFD of the intermediate node. The main different task is *IComp* which basically does the computation to take the average value of all received temperature and humidity values of all children, notes alarm conditions in received values, and activates alarms if needed by depositing alarm messages (*comp_al*) in the *comp_alarm* data storage.

13.9.2 Detailed Design and Implementation

The local monitoring of this node will be the same as the leaf and the intermediate node, so we will not repeat the functions of the *Temp*, *Humid*, and *Accel* tasks and data structures. The message structure defined below may be included in the data unit structure defined in Chap. 5 to be used for this application.

```
/****************************************************************
                   Cluster Message Structure
 ****************************************************************/
typedef struct {
        ushort node_id;
        ushort type;
        } alarm_type;

typedef struct {
        ushort cluster_id;
        double ctemp_ave;
        double chumid_ave;
        alarm_type alarm_cond[N_CLUS_NODE];
        } cluster_msg_t;
```

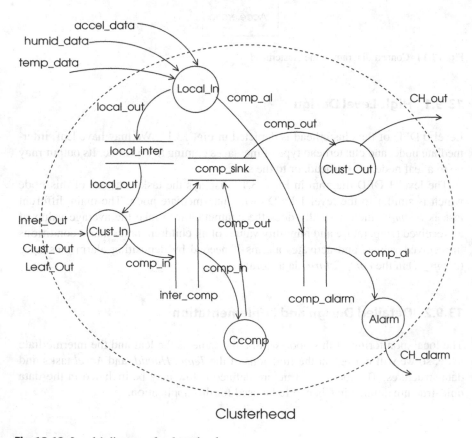

Clusterhead

Fig. 13.12 Level 1 diagram of a clusterhead

The front-end task called *Clust_In* collects data from all of the children of the node which may be local cluster node data or other data in the form of a cluster message. It works as a router by switching any out of cluster messages to the output task *Clust_Out* and inter-cluster data messages to the *Ccomp* task to be processed as in the intermediate node. We will omit the code for this task as it is very similar to the *Inter_In* task code of the intermediate node. The *Ccomp* task works differently than its counterpart *Imerge* task as shown below. It inputs all of the data and takes the average of the received values, forms a compact message containing these values along with the alarm conditions, and deposits this message in the mailbox of the *Clust_Out* task as shown below. Note that checking message type is similar to checking message types in the intermediate node so we can use the *check_interm* function here. The message type will be determined by this function either locally or by calling the *check_local* function.

```
/******************************************************************
                          Ccomp Task
 ******************************************************************/
data_unit_ptr_t data_pt, data_pt2, recvd_pt;
node_msg_ptr_t nddata_pt;
inter_msg_ptr_t intdata_pt;
clust_msgs_ptr_t clusdata_pt;
ushort mbox_id1=task_tab[current_tid].mailbox_id;
ushort mbox_id2=task_tab[System_Tab.DL_Out_id].mailbox_id;
double temp_tot=0.0, humid_tot=0.0;
ushort child_count=0, msg_count=0,i,n;


TASK Ccomp() {

  data_pt=get_data_unit(System_Tab.userpool1);
  nddata_pt=&(data_pt->data.node_msg);
  intdata_pt=&(data_pt->data.interm_msg);
  clusdata_pt=&(data_pt->data.clust_msg);

  while(TRUE) {
    delay_task(current_tid,System_Tab.LOCAL_DELAY);
    recvd_pt=recv_mailbox_wait(mbox_id1);
    check_interm(recvd_pt->TL_header.type, data_pt);
    if(child_count==task_tab[current_tid].n_children+1) {
     data_pt2=get_data_unit(System_Tab.userpool1);
     nddata_pt=&(data_pt->data.node_msg);
     for(i=0;i<msg_count;i++) {
      temp_tot=temp_tot+nddata_pt->temp.value;
      humid_tot=humid_tot+nddata_pt->humid.value;
      nddata_pt++;
     }
    data_pt2->data.cluster_msg.temp_ave=(double)temp_tot/msg_count;
    data_pt2->data.cluster_msg.humid_ave=(double)humid_tot/msg_count;
     data_pt2->data.cluster_msg.cluster_id=System_Tab.this_node;
     data_pt2->TL_header.type=CHEAD_MSG;
```

```
        data_pt2->TL_header.sender_id=System_Tab.this_node;
        data_pt2->MAC_header.sender_id=System_Tab.this_node;
        data_pt2->MAC_header.receiver_id=my_parent;
        send_mailbox_notwait(mbox_id2,data_pt2);
        data_pt=get_data_unit(System_Tab.userpool1);
        nddata_pt=&(data_pt->data.node_msg);
        intdata_pt=&(data_pt->data.interm_msg);
        child_count=msg_count=0;
        data_pt2=get_data_unit(System_Tab.userpool1);
    }
  }
}
```

13.10 The Sink

The sink is a node with advanced computational capabilities. It takes cluster information from clusterheads and displays this information on cluster basis in its monitor with highlighting the alarm conditions. It also provides a short report to a remote station over a gateway.

13.10.1 High-Level Design

Based on the requirements of the sink node, a context diagram can be drawn as in Fig. 13.13.

The Level 1 DFD of the sink is depicted in Fig. 13.14. Note that the sink does not do any local processing so we do not have this functionality at this node. The alarm conditions of nodes are already determined so there is no alarm checking either. Lastly, we assume the cluster data is displayed online as it is received; hence, the

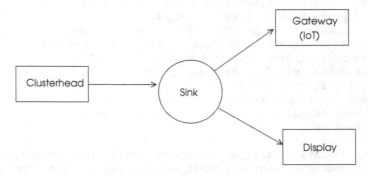

Fig. 13.13 Context diagram of the sink

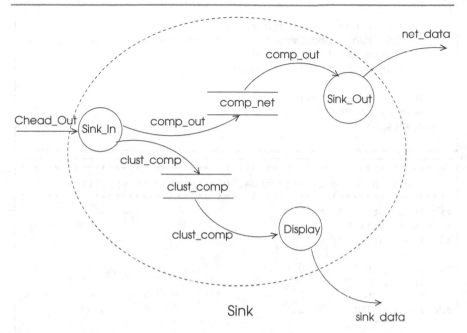

Fig. 13.14 Level 1 DFD of the sink

sink does not need to wait for the collection of data from its children. It can be seen that dividing the functions among various types of nodes has resulted in a simple sink design. Detailed design of the sink is similar to the design and coding of intermediate and clusterhead nodes and is left as a programming project (See Exercise 3).

13.11 Testing

The sample kernel DRTK will be used for simulation, and the inputs from the sensors will be realized by reading values from input files. We need to modify the data unit structure defined in Chap. 5. The transport layer header and MAC layer headers remain but the data area should be a union of the new types of messages introduced in this case study as follows. Tasks created are the coded ones only.

```
/******************************************************************
                    New Data Unit Structure
******************************************************************/
typedef struct *data_ptr_t{
    TL_header_t TL_header;
    MAC_header_t MAC_header;
    int type;
    int data[N_DATA];
    union data {
```

```
            sensor_msg_t sensor_msg;
            node_msg_t node_msg;
            interm_msg_t interm_msg;
            cluster_msg_t cluster_msg;
        }
    ushort MAC_trailer;
    data_ptr_t next;
}data_unit_t;
```

```
/****************************************************************
        main program at each node, clusterhead is specified
****************************************************************/
#include <pthread.h>
#include "drtk.h"
#include "drtk.c"

#define     LEAF        0
#define     INODE       1
#define     CH          2
#define     SINK        3

void main() {
    ushort my_id, my_parent;

    // select what to compile

    #ifndef SINK // make sensor devices, sink doesn't have them
        System_Tab.temp_devid=make_dev();
        System_Tab.humid_devid=make_dev();
        System_Tab.alarm_devid=make_dev();
        System_Tab.Temp_id=make_task(Temp,SYSTEM,1,NO);
        System_Tab.Humid_id=make_task(Humid,SYSTEM,1,NO);
        System_Tab.Accel_id=make_task(Accel_Int,SYSTEM,1,NO);
        System_Tab.Alarm_id=make_task(Leaf_Alarm,SYSTEM,1,NO);
        System_Tab.Leaf_In_id=make_task(Leaf_In,SYSTEM,1,NO);
        System_Tab.Leaf_Out_id=make_task(Leaf_Out,SYSTEM,1,NO);
    #endif

    #ifdef INODE
        System_Tab.Inter_In_id=make_task(Inter_In.SYSTEM,1,NO);
        System_Tab.Imerge_id=make_task(Imerge,SYSTEM,1,NO);
        System_Tab.Inter_Out_id=make_task(Inter_Out,SYSTEM,1,NO);
    #endif

    #ifdef CH
        System_Tab.Ccomp_id=make_task(Ccomp,SYSTEM,1,NO);
    #endif
    . . .
    Schedule();
}
```

13.12   Alternative Implementation with POSIX Threads

We will now consider an alternative detailed implementation method with POSIX threads without using the DRTK functions. Each task is realized with a POSIX thread in this case and the inter-thread communication method described in Chap. 4 can be used for task communications. We will just describe the changes to be made to the existing code described up to this point:

- Tasks are realized by the *pthread_create* POSIX call instead of DRTK *make_task* system call.
- The DRTK task communication routines are replaced by the simple procedures *write_fifo* and *read_fifo* described in Chap. 4.
- Delaying of tasks can be accomplished by using the *sleep* and *usleep* functions instead of DRTK *delay_task* function.
- The scheduling of POSIX threads is done by the system.

This approach may seem simpler than implementing DRTK; however, we need to have a POSIX interface typically in a UNIX-like system which would require significant memory capacity.

13.13   Chapter Notes

We described the design and implementation of a distributed real-time system as a case study in this chapter. Starting with the requirement specification and analysis, the decision was to employ a WSN for the system hardware regarding the cost, functionality, and ease of deployment. The coarse timing analysis showed there are no strict requirements. The second consideration was the hardware–software co-design and we saw that deciding to use a WSN resulted in the implementation of a spanning tree and clustering algorithm which we showed in detail.

The algorithm employed classifies the nodes of the WSN as leaf, intermediate, or clusterhead nodes. We then performed high-level design, detailed design, and coding for all of these node types and the sink computer using DFDs. The real-time operating system kernel to use is the sample kernel DRTK which provides the basic inter-task synchronization and communication primitives along with functions to create objects such as tasks, device control blocks, and to manage the network. The testing involves compiling the modules for each node and then executing. There are some modifications to be done to DRTK which were not shown in detail. These include adding few more fields to system table and a real application would require integrating the network drivers of the WSN nodes. Nevertheless, we think that this way of applying the design steps provides an insight into the implementation of a distributed real-time system.

13.14   Programming Exercises

1. Show how fault tolerance can be achieved while collecting data from children in the case of intermediate, clusterhead, and sink nodes, when one or more children fail to deliver its data.
2. Modify the code for the *Icomp* task of the intermediate node so that this task directly sends data to an upstream task over the network.
3. Perform detailed design of the sink software by forming all needed tasks.
4. *Team Project*: Realize the WSN monitoring system using POSIX threads and the inter-thread communication module.

Reference

1. Erciyes K (2018) Guide to graph algorithms: sequential, parallel and distributed. Springer Nature

Pseudocode Conventions

A

A.1 Introduction

In this part, the pseudocode conventions for writing an algorithm is presented. The conventions we use follow the modern programming guidelines and are similar to the ones used in [1,2]. Every algorithm has a name specified in its heading and each line of an algorithm is numbered to provide citation. The first part of an algorithm usually starts by its inputs. Blocks within the algorithm are shown by indentations. The pseudocode conventions adopted are described as data structures, control structures, and distributed algorithm structure as follows.

A.2 Data Structures

Expressions are built using constants, variables, and operators as in any functional programming language and yield a definite value. *Statements* are composed of expressions and are the main unit of executions. All statements are in the form of numbered lines. Declaring a variable is done as in languages like Pascal and C where its type precedes its label with possible initialization as follows:

$$\textbf{set of int} \quad neighbors \leftarrow \{\varnothing\}$$

Here, we declare a set called *neighbors* of a vertex in a graph each element of which is an integer. This set is initialized to $\{\varnothing\}$ (empty) value. The other commonly used variable types in the algorithms are *boolean* for boolean variables and *message types* for the possible types of messages. For assignment, we use $\leftarrow$ operator which shows that the value on the right is assigned to the variable in the left. For example, the statement:

$$a \leftarrow a + 1$$

© Springer Nature Switzerland AG 2019

K. Erciyes, *Distributed Real-Time Systems*, Computer Communications and Networks, https://doi.org/10.1007/978-3-030-22570-4_A

Table A.1 General algorithm conventions

Notation	Meaning
$x \leftarrow y$	Assignment
$=$	Comparison of equality
$\neq$	Comparison of inequality
true, *false*	Logical true and false
null	Nonexistence
$\triangleright$	Comment

Table A.2 Arithmetic and logical operators

Notation	Meaning
$\neg$	Logical negation
$\wedge$	Logical and
$\vee$	Logical or
$\oplus$	Logical exclusive-or
x/y	x divided by y
$x \cdot y$ or xy	Multiplication

increments the value of the integer variable a. Two or more statements in a line are separated by semicolons and comments are shown by $\triangleright$ symbol at the end of the line as follows:

$$1 : a \leftarrow 1; c \leftarrow a + 2; \qquad \triangleright \quad c \;\; is \;\; now \;\; 3$$

General algorithmic conventions are outlined in Table A.1.

Table A.2 summarizes the arithmetic and logical operators used in the text with their meanings.

Sets instead of arrays are frequently used to represent a collection of similar variables. Inclusion of an element u to a set S can be done as follows:

$$S \leftarrow S \bigcup \{u\}$$

and deletion of an element v from S is performed as follows:

$$S \leftarrow S \setminus \{v\}$$

Table A.3 shows the set operations used in the text with their meanings.

A.3 Control Structures

In the sequential operation, statements are executed consecutively. Branching to another statement can be done by *selection* described below.

Table A.3 Set operations

Notation	Meaning
$\|S\|$	Cardinality of S
$\varnothing$	Empty set
$u \in S$	u is a member of S
$S \bigcup R$	Union of S and R
$S \bigcap R$	Intersection of S and R
$S \setminus R$	Set subtraction
$S \subset R$	S is a proper subset of R
max/min S	Maximum/minimum value of the elements of S
max/min$\{\ldots\}$ S	Maximum/minimum value of a collection of values

Selection

Selection is performed using conditional statements which are implemented using *if-then-else* in the usual fashion and indentation is used to specify the blocks as shown in the example code segment below

---
Algorithm A.1 *if-then-else structure*

```
 1: if condition then                                          ▷ first check
 2:     statement1
 3:     if condition2 then                                     ▷ second (nested) if
 4:         statement2
 5:     end if                                                 ▷ end of second if
 6: else if condition3 then                                    ▷ else if of first if
 7:     statement3
 8: else
 9:     statement4
10: end if                                                     ▷ end of first if
```

In order to select from a number of branches, *case-of* construct is used. The expression within this construct should return a value which is checked against a number of constant values and the matching branch is taken as follows:

```
1. case expression of
2.     constant₁ : statement₁
3.     ⋮
4.     constantₙ : statementₙ
5. end case
```

Repetition

The main loops in accordance with the usual high level language syntax are the *for*, *while* and *loop* constructs. The *for-do* loop is used when the count of iterations can be evaluated before entering the loop as follows:

1. **for** $i \leftarrow 1$ **to** n **do**
2. $\vdots$
3. **end for**

The second form of this construct is the *for all* loop which arbitrarily selects an element from the set specified and iterates until all members of the set are processed as shown below where a set S with three elements and an empty set R are given and each element of S is copied to R iteratively.

1. $S \leftarrow \{3, 1, 5\}; R \leftarrow \varnothing$
2. **for all** $u \in S$ **do**
3. $R \leftarrow R \bigcup \{u\}$
4. **end for**

For the indefinite cases where the loop may not be entered at all, the *while-do* construct may be used where the boolean expression is evaluated and the loop is entered if this value is true as follows:

1. **while** *boolean expression* **do**
2.   *statement*
3. **end for**

A.4   Distributed Algorithm Structure

Distributed algorithms have significantly different structures than the sequential algorithms as their execution pattern are determined by the type of messages they receive from their neighbors. For this reason, the general distributed algorithm pseudocode usually includes a similar structure to the algorithm template shown in Algorithm A.2.

In this algorithm structure, there may be n types of messages and the type of action depends on the type of message received. For this example, the **while-do** loop executes as long as the value of the boolean variable *flag* evaluates to *true*. Typically, a message received by this node (i) at some point triggers an action which changes the value of the *flag* variable to *true* which then results in this termination of the loop. In another frequently used distributed algorithm structure, the **while-do** loop is executed forever and one or more of the actions should provide *exit* from this endless *while* loop as shown in Algorithm A.3.

Algorithm A.2 *Distributed Algorithm Structure 1*

1: **int** i, j ▷ i is this node; j is the sender of the current message
2: **while** $\neg flag$ **do**   ▷ all nodes execute the same code
3:　　**receive** $msg(j)$
4:　　**case** $msg(j).type$ **of**
5:　　　　　　$type\_1$: $Action\_1$
6:　　　　　　... : ...
7:　　　　　　$type\_n$: $Action\_n$
8:　　**if** $condition$ **then**
9:　　　$flag \leftarrow true$
10:　　**end if**
11: **end while**

Algorithm A.3 *Distributed Algorithm Structure 2*

1: **while** *forever* **do**
2:　　**receive** $msg(j)$
3:　　**case** $msg(j).type$ **of**
4:　　　　　　$type\_1$: $Action\_1$: **if** $condition_1$ **then exit**
5:　　　　　　... : ...
6:　　　　　　$type\_x$: $Action\_1$: **if** $condition_x$ **then exit**
7:　　　　　　$type\_n$: $Action\_n$
8: **end while**

The indefinite structure of this loop type makes it suitable for distributed algorithms where the type of message, in general, can not be determined beforehand.

Lower Kernel Functions

B

B.1 Data Unit Queue System Calls

```
/* data_unit_que.c*/

/*************************************************************
                check a data unit queue
*************************************************************/

int check_data_que(data_que_ptr_t dataque_pt) \{

   if (dataque_pt->front == NULL)
     return (EMPTY);
   return(FULL);
}


/*************************************************************
          enqueue a data unit to a queue
*************************************************************/

int enqueue_data_unit(data_que_ptr_t dataque_pt, data_unit_ptr_t
data_pt) {

    data_unit_ptr_t temp_pt;
    data_pt->next=NULL;
    if (dataque_pt->front != NULL) {
       temp_pt=dataque_pt->rear;
       temp_pt->next=data_pt;
       dataque_pt->rear=data_pt;
    }
    else
      dataque_pt>front=dataque_pt->rear=data_pt;
```

© Springer Nature Switzerland AG 2019

K. Erciyes, *Distributed Real-Time Systems*, Computer Communications
and Networks, https://doi.org/10.1007/978-3-030-22570-4_B

```
        return(DONE);
}

/***************************************************************
          dequeue a data unit from a queue
***************************************************************/

data_unit_ptr_t dequeue_data_unit(data_que_ptr_t dataque_pt) {

    data_unit_ptr_t data_pt;
    if (dataque_pt->front!=NULL) {
        data_pt=dataque_pt->front;
        dataque_pt->front=data_pt->next;
    }
    return(data_pt);
  }
  return(ERR_NOT_AV);
}
```

B.2   Task Queue System Calls

```
/* task_que.c */

/***************************************************************
                check a task queue
***************************************************************/

int check_task_que(task_que_ptr_t taskque_pt) {

    if (taskque_pt->front == NULL)
      return (EMPTY);
    return(FULL);
}

/***************************************************************
            enqueue a task to a task queue
***************************************************************/

int enqueue_task(task_que_ptr_t taskque_pt, task_ptr_t  task_pt){

    task_ptr_t temp_pt;
    task_pt->next=NULL;
    if (taskque_pt->front != NULL) {
       temp_pt=taskque_pt->rear;
       temp_pt->next=task_pt;
       taskque_pt->rear=task_pt;
    }
```

```
    else
      taskque_pt>front=taskque_pt->rear=task_pt;
    return(DONE);
}

/****************************************************************
            dequeue a task from a task queue
****************************************************************/
task_ptr_t dequeue_task(task_que_ptr_t taskque_pt) {

    task_ptr_t task_pt;
    if (taskque_pt->front!=NULL) {
       task_pt=taskque_pt->front;
       taskque_pt->front=task_pt->next;
       return(task_pt);
    }
    return(ERR_NOT_AV);
}

/****************************************************************
    insert a task to a task  queue according to priority
****************************************************************/

int insert_task(task_que_ptr_t taskque_pt, task_ptr_t task_pt) {

    task_ptr_t task_pt, temp_pt, previous_pt;

    if(task_pt->priority < taskque_pt ->front->priority) {
       temp_pt=taskque_pt.front;
       taskque_pt.front=task_pt;
       task_pt->next=temp_pt;
    }
    else {
       previous_pt=taskque_pt.front->next;
       while(task_pt-> priority >= previous_pt.priority) {
          previous_pt=temp_pt;
          temp_pt=temp_pt->next;
       }
       previous_pt.next=task_pt;
       task_pt->next=temp_pt;
    }
    return(DONE);
}

/****************************************************************
                insert a task to delta queue
****************************************************************/

int insert_delta_queue(ushort task_id, ushort n_ticks) {
    task_ptr_t task_pt, temp_pt, previous_pt, next_pt;
    task_queue_ptr_t taskque_pt;
    ushort total_delay=0;
```

```
    if (task_id < 0 || task_id >= System_Tab.N_TASK)
        return(ERR_RANGE);
    task_pt=&(task_tab[task_id]);
    task_pt->delay_time=n_ticks;
    taskque_pt=&delta_que;
    if (task_pt->delay_time < taskque_pt->front->delay_time) {
        temp_pt=taskque_pt.front;
        taskque_pt.front=task_pt;
        task_pt->next=temp_pt;
        temp_pt->delay_time=temp_pt->delay_time-task_pt
        ->delay_time;
        }
    else {
        previous_pt=taskque_pt.front;
        total_delay=taskque_pt->front->delay_time;
        while(task_pt->delay_time > total_delay) {
            previous_pt=next_pt;
            next_pt=previous_pt->next;
            if (next_pt==NULL) {
              next_pt->next=task_pt;
              task_pt->next=NULL;
              task_pt->delay=task_pt->delay-total_delay;
            }
            total_delay=total_delay+next_pt->delay_time;
        }
        previous_pt.next=task_pt;
        task_pt->next=next_pt;
        task_pt->delay=task_pt->delay-total_delay-next_pt->delay;
        next_pt->delay=next_pt->delay-task_pt->delay;
    }
    return(DONE);
}
```

References

1. Cormen TH, Leiserson CE, Rivest RL, Stein C (2001) Introduction to algorithms. MIT Press
2. Smed J, Hakonen H (2006) Algorithms and networking for computer games. Wiley. ISBN: 0-470-01812-7

Index

Printed in the United States
By Bookmasters